CENTRAL REGULATION OF THE PITUITARY-ADRENAL COMPLEX

STUDIES IN SOVIET SCIENCE

1973

Motile Muscle and Cell Models
N. I. Arronet

Densification of Metal Powders during Sintering
V. A. Ivensen

The Transuranium Elements
V. I. Goldanskii and S. M. Polikanov

Pathological Effects of Radio Waves
M. S. Tolgskaya and Z. V. Gordon

Gas-Chromatographic Analysis of Trace Impurities
V. G. Berezkin and V. S. Tatarinskii

A Configurational Model of Matter
G. V. Samsonov, I. F. Pryadko, and L. F. Pryadko

Complex Thermodynamic Systems
V. V. Sychev

Central Regulation of the Pituitary-Adrenal Complex
E. V. Naumenko

Crystallization Processes Under Hydrothermal Conditions
A. N. Lobachev

STUDIES IN SOVIET SCIENCE

CENTRAL REGULATION OF THE PITUITARY-ADRENAL COMPLEX

E.V. Naumenko

Institute of Cytology and Genetics
Siberian Branch, Academy of Sciences of the USSR
Novosibirsk, USSR

Translated from Russian by
Basil Haigh

CONSULTANTS BUREAU • NEW YORK-LONDON

Library of Congress Cataloging in Publication Data

Naumenko, Evgeniĭ Vladimirovich.
Central regulation of the pituitary-adrenal complex.

(Studies in Soviet science)
Translation of T͡sentral'nai͡a reguli͡at͡sii͡a gipofizarno-nadpochechnikovogo kompleksa.
Bibliography: p.
1. Pituitary body. 2. Adrenal glands—Innervation. I. Title. II. Series. [DNLM: 1. Pituitary-adrenal system—Physiology. WK510 N311t 1973]
QP188.P58N3513 599'.01'42 73-17250
ISBN 978-1-4684-1616-9 ISBN 978-1-4684-1614-5 (eBook)
DOI 10.1007/978-1-4684-1614-5

The original Russian text, published for the Institute of Physiology of the Siberian Branch of the Academy of Sciences of the USSR by Nauka Press in Leningrad in 1971, has been corrected by the authors for the present edition. This translation is published under an agreement with Mezhdunarodnaya Kniga, the Soviet book export agency.

TSENTRAL'NAYA REGULYATSIYA GIPOFIZARNO-NADPOCHECHNIKOVOGO KOMPLEKSA

E. V. Naumenko

ЦЕНТРАЛЬНАЯ РЕГУЛЯЦИЯ ГИПОФИЗАРНО-НАДПОЧЕЧНИКОВОГО КОМПЛЕКСА

Е. В. Науменко

A Division of Plenum Publishing Corporation
227 West 17th Street, New York, N.Y. 10011

United Kingdom edition published by Consultants Bureau, London
A Division of Plenum Publishing Company, Ltd.
Davis House (4th Floor), 8 Scrubs Lane, Harlesden, London, NW10 6SE, England

Preface

This monograph gives an up-to-date account of the original results of the author's research in the complex field of the central control over relationships which have become established in the course of evolution between the hypothalamus and the pituitary gland. The author has limited the scope of this research on logical grounds to the study of central chemically reactive structures in the regulation of the adrenal glands through the intermediary of the hypothalamo-hypophyseal system. Through the author's skill and expertise in the analysis of the extensive and sometimes complicated literature he has successfully undertaken a differential approach to the analysis of the concrete role of catecholamines, acetylcholine, and serotonin in relation to the endocrine system.

The vast range of experimental investigations and results of clinical observations published in the last few decades have revealed the humoral neurosecretory factors of the hypothalamus as important regulators of the principal functions of the pituitary gland and, through it, of the cyclic activity of the peripheral group of endocrine glands. Until recently the region of the cascade of regulatory influences preceding the hypothalamus was terra incognita. It is for this reason that Evgenii Vladimirovich Naumenko's monograph acquires its special importance for the neurophysiologist, for its author has striven, by the use of experimental methods, to ascertain in detail the nature of influences which for a long time were interpreted as the general dogmas of an abstract "nervism."*

*A theory of the unity of the organism and external environment in which the nervous system is the principal connecting link. The nervous system plays the principal role in determining the course of normal (adaptive) and pathological (protective) processes. (Translator's note.)

One original result of these investigations has been to obtain direct evidence that the higher levels of the brain have a direct influence on the localization of the system of serotoninergic structures located in them. The differentiation of the intimate relations between chemically reactive structures and the different zones of the hypothalamic region is a valuable achievement. No cholinergic or adrenergic structures capable of influencing the hypophyseal zone of the hypothalamus, secreting the corticotrophin-releasing factor, have been found actually in the brain itself. The corresponding stimulation must be secondary in this respect, for it is mediated through peripheral reactive structures. Cholinergic factors act in the same way. The author puts forward a soundly based hypothesis according to which certain neurons whose endings activate the secretion of corticotrophin-releasing substances in the hypothalamus are serotoninergic in nature. Participation of the serotoninergic structures of the brain in the direct central control over the hypothalamo-hypophyseo-adrenal system is thus stipulated for the first time.

The experimental results are illustrated by a number of lucid and informative schemes. The style of the description is perhaps a little laconic, and this may make it more difficult to digest by readers whose interests lie outside the field of neuroendocrinology. By way of compensation, however, this concentrated presentation of the facts, conclusions, and hypotheses enhances the value and interest of the book to the specialist.

A. A. Voitkevich

Obninsk, March 1968

Contents

Introduction 1

Chapter I. Role of Different Parts of the Brain as Regulators of the Pituitary-Adrenal System 5
Role of the hypothalamus 5
The hypothalamic factor stimulating ACTH secretion 6
Influence of different parts of the hypothalamus 9
The hypothalamus as a whole and as part of the central mechanism of regulation 21
The limbic system of the telencephalon 22
The amygdalar complex 23
The hippocampus 26
The septum 28
Other structures of the limbic system 29
The role of the mesencephalon 30
The tegmentum mesencephali 31
The mesencephalic reticular formation 32
Emotional factors and the pituitary-adrenocortical system 33

Chapter II. Adrenergic Structures and Regulation of Pituitary-Adrenal Function .. 39
Noradrenalin as a hypothetical mediator in the brain 41
Effect of catecholamines and substances exciting and blocking adrenergic structures on the pituitary-adrenocortical system 44
Catecholamines and adrenomimetics 44
Adrenolytics 50

Ways in which catecholamines influence the pituitary-adrenal complex 52
Peripheral catecholamines 52
Effects of adrenomimetics on guinea pigs with an intact brain 55
Action of adrenomimetics on guinea pigs after mesencephalic section 61
Role of brain catecholamines 65
Effect of local injection of noradrenalin into the brain on animals with an intact brain and with mesencephalic brain section 67

Chapter III. Cholinergic Structures and the Regulation of Function of the Pituitary-Adrenal System 75
Acetylcholine as a hypothetical mediator in the central nervous system 75
Action of acetylcholine and of substances exciting and blocking cholinergic structures on the pituitary-adrenocortical system 77
Pathways for the effect of acetylcholine on the pituitary-adrenal complex 78
Role of peripheral cholinergic structures 78
The role of central cholinergic structures 80
Effect of anticholinesterase preparations on guinea pigs with an intact brain 83
Action of anticholinesterase drugs after mesencephalic section 89
Effect of local intracerebral injection of carbachol on animals with an intact brain and after brain section 93

Chapter IV. Serotoninergic Structures and Regulation of Pituitary-Adrenal Function 97
Serotonin as a hypothetical mediator in the central nervous system 97
The adrenal cortex, stress, and serotonin 100
Adrenalectomy 100
Administration of adrenocortical hormones 101
Stress and the serotonin content in the body 102

Effect of serotonin on the pituitary-adrenal complex 104
Disturbance of the serotonin balance and the pituitary-adrenal complex 105
Changes in the rate of destruction 105
Disturbance of the ability of the tissues to retain serotonin 109
Changes in the intensity of synthesis 111
The central action of serotonin 116
Intraventricular injection of serotonin into animals with an intact brain 116
Injection of serotonin into the lateral ventricle after mesencephalic section 117
Localization of serotoninergic structures in the brain 118
Hypothalamus 118
Effect of 5-Hydroxytryptophan on the pituitary-adrenal system of animals with chronic deafferentation of the mediobasal hypothalamus 128
The limbic system of the telencephalon 129
Mesencephalon 135

Conclusion 139

Bibliography 157

Introduction

Many years have passed since Selye (1937) first described the nonspecific generalized response of the pituitary-adrenal system to stress, and the general acceptance that this response plays an important role in adaptation. Since those days, interest in the study of the complex mechanism responsible for the secretion of the adrenocorticotropic hormone (ACTH) by the anterior lobe of the pituitary gland has continued undiminished.

Advances in biochemistry and the introduction of new neuropharmacological research techniques have led to considerable progress in the field of endocrinology and, in particular, in the study of the central regulation of the hypophyseo-adrenocortical system. Recent investigations have yielded facts of fundamental importance: the role of the portal circulation of the pituitary for its adrenocortical function has been demonstrated, the hypothalamic substance (corticotropin-releasing factor) which stimulates the secretion of ACTH has been discovered and identified, the leading role of the hypothalamus in the control of hypophyseo-adrenal activity has been shown, and the nervous centers of the hypothalamus concerned with this process have been found. In the last decade a study has been started on the role of other parts of the central nervous system in anterior pituitary control. Attention has been concentrated on the mesencephalon and the limbic system of the telencephalon – structures anatomically and functionally connected directly with the hypothalamus.

Nevertheless current views on the localization of the central integrative mechanism controlling the pituitary-adrenal complex are still inadequate to explain the principles governing its work. In particular, we need to know which biologically active substances transmit nervous impulses to the hypophyseal zone of the hypothalamus and the pathways which these impulses follow. It is only by the solution of these complex problems, part of the greater problem of brain activity in general, that it will be possible to understand the central mechanisms controlling the hypophyseo-adrenocortical system.

In the modern view the chemical transmission of nervous impulses in synapses is a means of influencing the excitability of the postsynaptic cells. The mediator liberated at the ends of the nerve fibers reacts with the sensitive biochemical structures of the innervated cells. It is in this way that the nervous impulse is transmitted from neuron to neuron and from neuron to effector.

Of the many different biologically active principles found in the brain and capable of influencing chemically sensitive structures as neurohormones or mediators, those which have attracted most attention are substances whose mediator role has been demonstrated at the periphery (noradrenalin and acetylcholine), and also serotonin, whose role in the biological reactions taking place in the body has been confirmed recently by intensive research.*

The study of noradrenalin, acetylcholine, and serotonin as hypothetical mediators in the central nervous system has become increasingly promising with each successive year. This is clear from the determination of the localization of monoamines and acetylcholine in synaptic nerve endings, the discovery of the specific mechanisms of absorption, synthesis, and breakdown of these principles, and the discovery of nerve cells responding differently to local microinjection of acetylcholine, noradrenalin, and serotonin and thus containing biochemical structures with selective sensitivity toward these substances.

Meanwhile the possibility cannot be ruled out that the biogenic amines and acetylcholine may act as modulators of nervous impulses, i.e., in Florey's (1976) definition, as compounds formed in the cell, but not in the synapse, which influence the excitability of

*The role of biogenic amines and acetylcholine in the activity of the nervous system is examined in the surveys of Bloom and Giarman (1968), Michaelson (1967), and Votava (1967).

nerve cells and which are a normal component of the regulatory mechanisms controlling the activity of the nervous system.

Under the influence of modulators, changes can be brought about in the nervous system in parameters such as the quantity of mediator liberated presynaptically, the time of its secretion, the quantity of mediator necessary to react with the subsynaptic receptor, the time required for the change of membrane permeability evoked by the mediator, the degree of the change in permeability, the threshold of excitation of the postsynaptic cell, and so on (Florey, 1967).

There is no question now of the ability of noradrenalin, acetylcholine, and serotonin to stimulate the pituitary-adrenal system. Changes in the content and rate of metabolism of these substances in the brain in various situations of stress likewise have been demonstrated. There are thus weighty grounds for postulating the role of adrenergic, cholinergic, and serotoninergic structures of the central nervous system in neurohumoral regulation of the hypothalamo-hypophyseo-adrenal complex.

However, the role of the central chemically reactive structures has so far received very little study from this point of view, and, so far as the writer is aware, no general survey of the subject has yet been published. In this book an attempt will therefore be made to classify the corresponding material on the basis of an analysis of the literature and the results of his own experiments. The author will not, of course, try to give an exhaustive picture of the neurohumoral mechanisms of the brain concerned in the regulation of the pituitary-adrenal system. This book is devoted to only one aspect of neuroendocrinology: the role of the most probable mediators or modulators, namely noradrenalin, acetylcholine, and serotonin, in the central regulation of the hypothalamo-hypophyseo-adrenal complex.

Chapter I

Role of Different Parts of the Brain as Regulators of the Pituitary-Adrenal System

Role of the Hypothalamus

The hypothalamus occupies a special place among the structures of the brain for it contains the centers which control the hormonal functions of the anterior lobe of the pituitary gland. Because of its "geographical" position this phylogenetically old region of the brain is a part of two systems of structures: the limbic formations of the telencephalon and the hypothalamus, on the one hand, and the hypothalamus and mesencephalon on the other (Nauta, 1956, 1960). If the hypothalamic region is looked at from this point of view, it can be considered to be the nodal point of a wide nerve chain stretching from the cerebral hemispheres caudally to the lower boundary of the mesencephalon. Its functional state is under the constant influence of the centers of the telencephalic limbic system and the mesencephalic reticular formation (Harris, 1956, 1958; Gloor, 1956; Green, 1956, 1958; Nauta, 1956, 1958, 1960; Lundberg, 1960; Grashchenkov, 1963; Allikmets, 1964; Tonkikh, 1968).

In 1950 it was postulated that the hypothalamus contains a nervous mechanism controlling the secretion of ACTH (De Groot and Harris, 1950). These same workers showed that electrical stimulation of the thalamus and mammillary region in rabbits evokes lymphocytopenia, indicating increased secretion of ACTH, while destruction of the same region considerably weakens this response induced by stress. They suggested that nerve endings in the hypothalamus secrete a chemical mediator which enters the portal system of blood vessels which form a primary plexus in the median eminence [for details about the portal circulation of the pituitary gland the monographs of Harris (1955) and Szentagothai et al. (1965) must be consulted]. The substance passes along the portal vessels to reach the anterior lobe of the pituitary where it exerts a specific effect on the secretory elements producing ACTH.

This hypothesis lay at the basis of modern views regarding the primary role of the hypothalamus in regulation of the pituitary-adrenal complex. The wealth of experimental evidence now available suggests that there is a single hypothalamo-hypophyseo-adrenal system, a conclusion which confirms the close connection between the hypothalamus and the hypophyseo-adrenocortical system. The relationship of the hypothalamus to this system is not exceptional, for in modern endocrinology new facts are constantly being discovered to show that similar close relations exist between the hypothalamus and other endocrine systems: pituitary–thyroid gland, pituitary–gonads (Harris, 1955; Aleshin, 1960, 1964; Voitkevich, 1960; Szentagothai et al., 1965; Redding et al., 1966).

From the beginning of the 1950s details began to be obtained of the role of individual areas and nuclei of the hypothalamus in the regulation of hypophyseo-adrenocortical function, and this was followed by the intensive study of the nature and site of formation in the hypothalamus of the humoral principle stimulating ACTH secretion.

The Hypothalamic Factor Stimulating ACTH Secretion

In 1955 Saffran et al. (1955) and Guillemin and Rosenberg (1955) showed independently that extracts of the hypothalamus and posterior lobe of the pituitary stimulate the secretion of adrenocorticotrophic hormone. At Saffran's suggestion, the substance, which was subsequently isolated from hypothalamic extracts, was called corticotrophin-releasing factor (CRF).* In early investigations attempts were made to identify CRF with biologically active substances whose mediator role had already been established or postulated. However, it became evident that this substance is neither

*Another neurohumoral principle hypothetically present in the hind brain is known as hind brain factor (HBF) which, like CRF, stimulates the adrenocorticotropic function of the pituitary (Egdahl, 1959, 1960a, 1961a,b). However, later experiments rendered this hypothesis untenable (Egdahl, 1962; Matsuda et al., 1963; Wise et al., 1963, 1964).

adrenalin, acetylcholine, serotonin, nor histamine (Guillemin, 1955, 1957; Guillemin et al., 1957). Attempts to identify CRF with vasopressin and oxytocin also were doomed to disappointment. Although a polypeptide, very similar to them in composition and its chemical properties, corticotrophin-releasing factor is not identical with these hormones (Saffran et al., 1955; Guillemin, 1962; Andersen and Egdahl, 1964).

Later, Schally's group (Schally et al., 1960; Schally and Guillemin, 1960) produced a mass of evidence to show the existence of two different corticotrophin-releasing factors (α- and β-CRF), and they showed that the former has weaker activity than the latter as regards the stimulation of ACTH secretion. Unfortunately β-CRF has been isolated only in very small amounts, making the study of its chemical and physiological properties difficult. By the use of chromatographic methods, the α-CRF fraction was subsequently subdivided into two polypeptides, α_1-CRF and α_2-CRF, which were isolated in a pure form (Guillemin et al., 1962; Schalli et al., 1962,a,b; Guillemin, 1962), and their amino acid sequence has been established.

Besides the intensive study of the structure and chemical properties of CRF, its possible site of formation in the hypothalamus has also been sought. This task has proved to be much more difficult than had been expected, and even now it has not been completed. There are two fundamentally different points of view regarding the site of formation of CRF in the hypothalamus at the present time. One associates the production of CRF with the parvocellular nuclei of the ventral hypothalamus, while the other ascribes the leading role to the magnocellular neurosecretory nuclei. The first view, which has received the widest support, is defended in particular by Hungarian workers (Szentagothai et al., 1965). They conclude from combined morphological and physiological investigations that the site of formation of the releasing factors, including CRF, is the medial basal hypothalamus. This zone, described as the hypophyseotrophic zone, is made of the parvocellular nuclei, including the arcuate, anterior periventricular, ventromedial, and dorsomedial nuclei, and some small cells of the retrochiasmal region. Szentagothai et al. (1965) conclude that releasing factors cannot be ordinary synaptic mediators, and they regard them as a neurosecretory substance formed in the perikaryons of cells of the tubero-infundibular nuclei, which enters the portal circulation to reach the anterior lobe of the pituitary gland. While, at the same time, they deny any anatomical connection between the hypothalamo-neurohypophyseal system and the portal vessels, they consider that the neurosecretory Gomori-positive substance of the magnocellular nuclei (supraoptic and paraventricular) bears no relationship to hypothalamic factors activating the trophic functions of the adenohypophysis. Basically the same point of view is held in the Soviet Union by Aleshin (1964, 1971).

A powerful argument supporting the leading of role of the hypophyseotrophic zone in the formation of CRF is given by the results of experiments on hypophysectomized rats, in which the disturbances of function of the hypophyseo-adrenal system disappeared completely only when the pituitary gland taken from a donor was implanted into the median eminence (Szentagothai et al., 1965).

By the use of the technique of deafferentation of the medial basal hypothalamus, brilliantly developed by Halász (Halász and Pupp, 1965), new and valuable evidence confirming the view expressed above was obtained (Halász, 1969; Rethelyi and Halász, 1970). After complete deafferentation, when the hypophyseotrophic zone no longer received nervous influences from the rest of the brain, the hypothalamus was able to maintain almost the entire functional activity of the hypophyseo-adrenal system, unlike in animals after destruction of the median eminence. Both the diencephalon if completely isolated from external nervous influences (Ondo and Kitay, 1972) and the medial basal hypothalamus after complete deafferentation (Feldman et al., 1970; Makara et al., 1970) continue to respond to a wide range of stressors reaching the tissues of the isolated zone by the humoral route. Deafferentation of the hypothalamus in newborn guinea pigs is not followed by any disturbance of the development of CRF formation, for 85-95 days later adrenal activity remains undisturbed (Joseph et al., 1969).

Meanwhile, after isolation of the hypophyseotrophic zone the initial level of corticosteroids in the blood plasma (Halász et al., 1967a) and the ACTH concentration in the pituitary (Halász et al., 1967b) rise while the diurnal rhythm of ACTH secretion disappears (Halász et al., 1967a,b; Halász, 1969; Palka et al., 1969). This shows that the hypophyseotrophic zone is not completely autonomous but is under the control of other parts of the brain.

A completely different view of the role of the hypothalamus

in the regulation of the trophic functions of the adenohypophysis is held by workers who have used certain special histochemical and electron-microscopic methods (Scharrer and Scharrer, 1963; Polenov, 1968). They consider that the factors stimulating release of the adenohypophyseal trophic hormones are neurohormones contained in the Gomori-positive granules of neurosecretion and produced by the magnocellular hypothalamic (supraoptic and paraventricular) nuclei. Neurosecretory products pass from these nuclei simultaneously into the posterior lobe of the pituitary and the median eminence. The role of the parvocellular nuclei of the hypophyseotrophic zone of the hypothalamus, according to these views, is secondary or accessory. The parvocellular nuclei of the medial basal hypothalamus, it is assumed, produce neurotransmitters and not releasing factors. These transmitters, liberated at the sites of the axo-vasal synapses, affect the state of the capillary walls of the primary portal system (Scharrer, 1965). Control over the secretion and transport of peptide neurohormones contained in the Gomori-positive substance and carrying factors regulating the secretion of the adenohypophyseal trophic hormones is effected by mediator influences through the endings of the tubero-hypophyseal system (Polenov, 1968).

The situation can accordingly be summed up by saying that the first humoral link in the chain of mechanisms responsible for secretion of adrenal cortical hormones is located in the hypothalamus, that the chemical nature of the corticotrophin-releasing factor has been studied to a certain extent, and that intensive work is in progress to determine its localization. However, with no personal experience of the study of these problems the author can only refer the reader to the special literature.

Full details about the nature, properties, site of formation, and mechanism of action of CRF can be found in the surveys and monographs by Guillemin (1956, 1962), Sayers (1960), Eskin (1960), Voitkevich (1961), Fortier (1962), Venning (1965), Szentagothai et al. (1965), McCann and Dhariwal (1966), Burgus and Guillemin (1970), Aleshin (1971), and Yates et al. (1971).

Influence of Different Parts of the Hypothalamus

Besides studies of the humoral factor regulating ACTH secretion, much attention has also been paid to the study of the nervous components of regulation. Usually research of this type has been carried out on various species of mammals, using two techniques: electrical stimulation by means of permanently implanted electrodes or destruction of localized areas of the hypothalamus.

The Central Hypothalamus. It has been shown conclusively that the tuber cinereum, especially its central part (the median eminence), is connected with the function of the hypophyseo-adrenocortical system.

Electrolytic destruction of the median eminence is followed by a decrease in weight of the adrenals and inhibition of compensatory hypertrophy of the opposite gland after removal of one adrenal (Ganong and Hume, 1954; Moll, 1959, 1960; De Wied, 1961), by a fall in the blood levels of ACTH and corticosteroids (Hume, 1953; Hume and Nelson, 1955; Hume, 1958; Hume and Egdahl, 1959; Krieger and Wagman, 1961; Saba et al., 1962a,b), by blocking or diminution of the response of the hypophyseo-adrenal system to stress (McCann, 1952, 1953; Porter, 1954; Hume and Nelson, 1955; Bouman et al., 1957; Daily and Ganong, 1958; Hume, 1958; Hume and Egdahl, 1959; McCann and Haberland, 1960; Krieger and Wagman, 1961; Way et al., 1962), by a marked inhibition of corticosteroid formation in the adrenals after stress in experiments in vitro (de Wied, 1961) and, finally, by loss of the diurnal rhythm of corticosteroid secretion (Saba et al., 1963; Tronchetti et al., 1964; Slusher, 1964).

The disturbance of activity of the hypophyseo-adrenal system after destruction of the median eminence develops gradually: after 48 h its response to stress is blocked, but the ACTH concentration in the pituitary may still be unchanged, whereas after 3 weeks or more, not only is the function of the adrenal cortex blocked, but in response to stress the ACTH content in the pituitary is reduced by 50% (McCann and Haberland, 1960). Six weeks after destruction of the median eminence, secretion not only of hydrocortisone and corticosterone, but also of 11-deoxycortisol and aldosterone, is inhibited (Ganong et al., 1961).

However, the results of experiments to study the function of the hypophyseo-adrenal system in a resting state or during stress after destruction of the median eminence must be interpreted with caution. The reason is that the pituitary portal circulation is affected to a greater or lesser degree depending on the extent of the lesion, so that the possibility of development of atrophic changes in the adenohypophysis and the development of some degree of hypopituitarism consequently cannot be ruled out. For this reason no conclusions regarding the specific role of the nervous structures of the hypothalamus in the response of the pituitary-adrenal complex under physiological or extremal conditions can be drawn purely on the basis of experiments involving destruction of the median eminence.

Experiments to study the role of individual nuclei of the central hypothalamus are evidently more demonstrative. Their value can be judged indirectly from experiments in which diphtheria toxin is injected into guinea pigs. Twenty hours after injection of diphtheria toxin into the control animals, foci of hemorrhagic necrosis developed in the adrenal cortex, but this was not observed after destruction of the ventromedial and dorsomedial nuclear groups (Schmid et al., 1957).

More direct proof of the participation of these nervous centers in the regulation of adrenocortical function was obtained by Tonutti (1961). Destruction of the ventromedial and dorsomedial nuclei blocked the excretion of 17-hydroxycorticosteroids by the kidneys after administration of diphtheria toxin to the guinea pigs. Tonutti considers that against the background of destruction of the central group of nuclei, inhibition of the response of the pituitary-adrenal system is more complete than after destruction of the median eminence. Destruction of the above-mentioned nervous structures in rabbits completely blocked the lymphocytopenic response to injection of formalin (Ahren, 1962). Experiments in which methods of determining ACTH in the pituitary and corticosteroids in the peripheral blood confirmed these findings. Destruction of the arcuate and ventromedial nuclei in rats blocked the response of the pituitary-adrenal system to stress (Usher et al., 1967). These results are in agreement with those of experiments in which stimulation of the central hypothalamus led to stimulation of the function of the pituitary-adrenal system in dogs, cats, rabbits, and mice (Katsuki, 1961).

At the same time, it has recently been reported that electrical stimulation of the infundibular nuclei in cats is not accompanied by any increase in the ACTH level in the peripheral blood plasma, whereas analogous stimulation of the lateral hypothalamus at the level of the mammillary bodies causes an increase in the concentration of this hormone (Redgate, 1970). If all the available information on the role of the median eminence and lateral hypothalamus in the regulation of the pituitary-adrenal system is taken into consideration, it is clear that repetition of these experiments is required.

The Posterior Portion of the Hypothalamus. Experiments involving destruction, like those involving stimulation of the posterior part of the hypothalamus, in the hands of most investigators have yielded results which indicate the great importance of this region in the regulation of the pituitary-adrenal complex. De Groot and Harris (1950) showed originally that electrical stimulation of the posterior part of the tuber cinereum or mammillary bodies in rabbits induces a lymphocytopenic response, whereas destruction of these areas blocks the appearance of this phenomenon in response to stress. Their results were subsequently confirmed by many other workers using different criteria for assessing pituitary and adrenal function.

Electrical stimulation of the mammillary region in cats (Porter, 1953a,b; Endröczi and Lissak, 1963), monkeys (Porter, 1953a, 1954), rats (Endröczi et al., 1956), mice and rabbits (Katsuki, 1961), and dogs (Goldfien and Ganong, 1962; Kovach et al., 1970) is accompanied by an increase in adrenocortical activity. In dogs it has been shown that the posterior hypothalamus contains zones which respond differently to electrical stimulation: stimulation of the inferoposterior zones causes an increase in the rate of corticosteroid secretion whereas similar stimulation of the superoposterior part of the hypothalamus leads to inhibition of secretion after an initial transient increase (Suzuki et al., 1960). Meanwhile, in monkeys electrical stimulation of the medial part of the superoposterior hypothalamus is accompanied by an activating effect on the pituitary-adrenal system (Smith and Root, 1971).

The whole of the ventromedial hypothalamus evidently contains neural elements whose role is to stimulate the function of the pituitary-adrenal system, for electrical stimulation through electrodes implanted into the hypothalamus between the optic chiasma and mammillary bodies, and including the latter, increases corticosteroid secretion in monkeys (Mason, 1956).

In response to electrical stimulation of different zones of the hypothalamus in dogs, three possible types of secretion of 17-hydroxycorticosteroids are found: 1) after stimulation of the anterior hypothalamus an increase in the concentration of corticosteroids in the blood flowing from the adrenal gland begins quickly and persists for a long time; 2) during stimulation of the central hypothalamus, activation of the adrenal cortex appears quickly but is transient in duration; and finally 3) stimulation of the posterior hypothalamus evokes a slow increase in corticosteroid secretion which persists for a long time (Katsuki, 1961).

On the other hand, stimulation of the posterior hypothalamus and, in particular, of its caudal part inhibits or blocks the response of the pituitary-adrenal system to stress in cats (Porter, 1953b), rabbits (Harris, 1955), and monkeys (Porter, 1954). After local bilateral destruction of the posterior and superoposterior parts of the mammillary bodies in rabbits, the lymphocytopenic response to formalin stress is completely blocked, whereas injury to the anterior part of the mammillary bodies and premammillary region does not alter the lymphocytopenic response by comparison with that observed in animals with an intact hypothalamus (Ahren, 1962).

Some workers deny that the neural centers of the posterior hypothalamus play any part in the response of the adenohypophysis to stress. For example, McCann (1953) and Bouman et al. (1957) in rats, and Hume (1958) in dogs, found no changes in the response of the pituitary-adrenal complex to a stimulus after destruction of the mammillary bodies.

After destruction of these formations in dogs the resting secretion of corticosteroids remained at the normal level (Hume, 1958), and in cats electrical stimulation of the posterior hypothalamus, including the mammillary bodies, had no activating effect on the pituitary (Anand and Dua, 1955). The possible reasons for the differences between experimental results obtained by different workers will be discussed again below.

Anterior Hypothalamus. Although most investigators recognize the great importance of the median eminence and the central and posterior regions of the hypothalamus in the regulation of the activity of the pituitary-adrenal system, the data on the role of the anterior hypothalamus are more contradictory.

Using the eosinophil test in their early experiments, and later determining the blood levels of ACTH and corticosteroids, Hume et al. (Hume, 1949, 1953, 1958; Hume and Wittenstein, 1950; Hume and Nelson, 1955; Hume and Jackson, 1959) showed in chronic experiments on dogs that electrical coagulation of the anterior zone of the median eminence and of part of the anterior hypothalamus immediately posterior to the optic chiasma is followed by a decrease in the function of the pituitary-adrenal system. Exposure to stressors 1 month after such an operation no longer evokes the stimulation of the adrenal cortex which is usually observed in animals with an intact hypothalamus.

The depression of adrenocortical function after electrical coagulation of the anterior hypothalamus was confirmed by other investigators in experiments on rats (Bouman et al., 1957; Porter, 1963) and rabbits (Ahren, 1962). This applies primarily to destruction involving an area located immediately posterior to the optic chiasma (Brodish and Long, 1962) and the region between the crura of the fornix at the level of the paraventricular nuclei (Moll et al., 1961).

These experiments are in good agreement with the results of those using electrical stimulation of the anterior hypothalamus by means of permanently implanted electrodes. For example, after electrical stimulation of the anterior hypothalamus immediate ly above the optic chiasma or posterior to it, in monkeys the blood concentration of 17-hydroxycorticosteroids rises (Mason, 1958). Similar results were obtained in cats (Anand and Dua, 1955) and dogs (Katsuki, 1961; Kovach et al., 1970).

Repeated electrical stimulation of the preoptic zone, the suprachiasmatic area, the arcuate nucleus, or the anterior part of the median eminence in rats leads to an increase in weight of the adrenals and a tendency for the blood corticosteroid level to rise (Snyder and D'Angleo, 1963; D'Angelo et al., 1964). D'Angelo and Young (1966) showed once again that similar stimulation is followed by an increase in the corticosteroid concentration in the adrenals and blood of rats. However, this response does not take place after destruction of the median eminence.

However, Endröczi (1961) found that destruction of the anterior hypothalamus, including the region between the paraventricular nuclei, optic chiasma, and infundibulum, and extending to part of the preoptic region in rats leads to increased secretion of corticosteroids from the adrenals into the blood. The results of his experiments are in agreement with those in which an increase in the weight of the adrenals was found in rats after local lesions in the anteromedial portion of the hypothalamus or in an area lying between the paraventricular nuclei and the optic chiasma (Szentagothai et al., 1965).

It has also been shown that although electrolytic destruction of foci in the anterior hypothalamus of rats blocks the increase in the ACTH concentration in the pituitary against the background of stress, a definite increase in the corticosterone level in the peripheral blood is nevertheless observed under those conditions. The workers concerned conclude from these results that the lesions produced block ACTH synthesis only, and not its secretion, and they postulate the existence of two different mechanisms for synthesis and secretion of ACTH (Usher et al., 1967).

Meanwhile, according to other reports neither electrical stimulation nor destruction in the region of the anterior hypothalamus causes any change in pituitary-adrenal activity. Such observations have been made principally in connection with the supraoptic and paraventricular nuclei which, according to many investigators (Scharrer and Scharrer, 1954, 1963; Polenov, 1958, 1962, 1968; Voitkevich, 1960, 1961, 1964), belong to the hypothalamo-neurohypophyseal secretory system. De Groot and Harris (1950) observed absence of lymphocytopenia in rabbits after stimulation of the supraoptico-hypophyseal tract in the region of the anterior part of the median eminence. Electrical stimulation of the anterior group of nuclei (supraoptic and paraventricular) did not change the response of the pituitary-adrenal system in rats (Endröczi et al., 1956) and cats (Endröczi and Lissak, 1963), while their destruction in rats (McCann, 1953), cats (Laquer et al., 1955), and monkeys (Porter, 1954) was followed by the usual response to stress.

The Lateral Portion. Although Porter (1954) considered that destruction of the lateral part of the tuber cinereum and lateral part of the posterior hypothalamus causes eosinopenia in monkeys, subsequent experiments did not confirm his findings. By the use of a similar test on unanesthetized cats Anand and Dua (1955) found no change in adrenocortical function after stimulation of the lateral part of the tuber cinereum. Experiments to determine the rate of secretion of corticosteroids into the venous blood flowing from the adrenals in cats also show that electrical stimulation of the lateral part of the tuber cinereum and lateral part of the posterior hypothalamus does not change the response of the pituitary-adrenal complex (Endröczi and Lissak, 1963). The lateral zones of the hypothalamus are evidently unconnected with regulation of ACTH secretion. In any case these zones do not contain nervous mechanisms responsible for the increase in function activity of the pituitary-adrenal system.

The suggestion has been made that inhibitory mechanisms are present in the anterolateral zones of the hypothalamus, for stimulation depresses the adrenocorticotrophic function of the pituitary (Endröczi and Lissak, 1963).

The rostral zones of the anterior hypothalamus continue into the preoptic region, which also contains nervous structures influencing the pituitary-adrenal system.

The Preoptic Region. The preoptic region, which some investigators include with the hypothalamus, but others with the telencephalon,* is believed to contain nervous structures with both inhibitory and stimulating effects on the anterior lobe of the pituitary and adrenal glands.

Electrical stimulation of this region in monkeys causes a transient rise in the hydrocortisone level in the peripheral blood followed by a prolonged fall (Mason et al., 1960). An equally transient (for 15 min) increase in the corticosteroid concentration was found in dogs after acute local cooling of the preoptic region (Chowers et al., 1964).

Electrical stimulation of the preoptic region in cats reduces the rate of secretion of corticosteroids into the blood flowing from the adrenals (Endröczi and Lissak, 1963) and lowers the blood ACTH concentration (Slusher and Hyde, 1966).

*There is disagreement regarding both the origin of the preoptic zone and its anatomical affiliation. Craigie (1925) considers that this region belongs to the central telencephalon, Kuhlenbeck (1954) subdivides it into telencephalic and diencephalic parts, while Riley (1960) places it at the point where the telencephalon joins the diencephalon, which is anatomically closely connected with the hypothalamus. However, whatever the origin of the preoptic region, it provides anatomical and function connections between the limbic structures of the telencephalon and the anterior hypothalamus (Zeman and Innes, 1963).

Stimulation of neurons located in the lateral part of the preoptic region is evidently accompanied by stimulation of the adrenal cortex, while stimulation of neurons in the medial part has the opposite effect. Only 15 min after interrupted stimulation of the medial part of the preoptic region in cats for 5 min the rate of corticosteroid secretion was reduced. The inhibition of secretion of the hormone from the adrenal cortex is progressive and reaches its maximum by the first hour (Yoshio, 1964). Similar results were obtained by McHugh et al. (1966) working with monkeys. On the other hand, autostimulation of the lateral zones of the preoptic region through permanently implanted electrodes led to an increase in the concentration of corticosteroids in the blood and brain of monkeys (McHugh et al., 1966). However, according to Suzuki et al. (1960) electrical stimulation of the lateral zones in dogs inhibited adrenocortical function. Admittedly these workers cite only one case.

The facts described above indicate that nearly all regions of the hypothalamus, with the exception of its lateral zones, essentially contain neural elements with the property of influencing the function of the pituitary-adrenal system. The ability of the nervous structures of the hypothalamus, on excitation, to stimulate or inhibit pituitary-adrenal activity has led to the conclusion that the hypothalamic region contains nervous mechanisms concerned with the regulation of the pituitary-adrenal complex. However, the results obtained by many investigators, not to mention their interpretation, remain contradictory. Because of the very large number of investigations, differences in the techniques used and in the species of experimental animal, in the nature of the stimuli used and their parameters, and so on, it is impossible to pinpoint the actual cause of the contradictions in each individual case. All that can be done is to indicate in their general features the factors which can influence the results of the investigations and, consequently, the conclusions derived from them. I ought perhaps to say that most of the causes discussed below could be responsible for the contradictions in the results obtained by the study not only of the role of the hypothalamus, but also of other parts of the brain.

Possible Causes of Contradiction in the Experimental Results. The important role of the hypothalamus in the regulation of physiological systems and the wide variety of its effects make it difficult to interpret changes in the function of the endocrine glands after injury to the hypothalamus or its electrical stimulation. Changes in the activity of the pituitary-adrenal system are frequently secondary, and they cannot be regarded as the result of direct action on the hypothalamus. The problem is complicated still further by the close interaction taking place between the various endocrine processes, so that disturbances of one of them may be reflected in others (Szentagothai et al., 1965). The use of different indices as criteria of the activity of the pituitary-adrenal complex makes evaluation of the results extremely difficult. Conclusions based on experiments in which the functions of the adrenal cortex are judged on the basis of indirect tests must be looked at very critically. For example, one of the most widely used techniques in the 1950s, that based on counting the eosinophils (Thorn and Forsham, 1959), by no means always characterizes the state of the pituitary-adrenal system as a whole correctly (Glin, 1957; Svirchevskaya, 1958; Kudryavtsev, 1959; Zaitsev, 1959; Schäfer, 1961). Dissociation frequently arises between the decrease in the number of eosinophils and the blood corticosteroid level (Glin, 1957; Naumenko and Davydova, 1965; Foss et al., 1971). The decrease in the white blood cell count in stress takes place when the function of the adrenal is depressed or after adrenalectomy (Kark and Muehrcke, 1952; Kolpakov et al., 1962a). Finally, the white blood cell (including eosinophil) count in stress can be changed by hypothalamo-hypophyseal block (Lenguel, 1964) by the extrahypothalamo-hypophyseal route (Schweizer, 1953; Kolpakov et al., 1962b).

Similar discrepancies are found between the blood level or rate of secretion of the adrenocortical hormones and changes in the ascorbic acid content of the adrenal cortex against the background of stress (Slusher, 1958; Gervazieva, 1962).

The effect of injury to the hypothalamus or its electrical stimulation on the weight and histological structure of the adrenals likewise is not particularly convincing: some results show that neither the weight of the adrenals nor their morphological study can provide a reliable criterion for assessment of the function of these glands (Sanzari et al., 1965; Szentagothai et al., 1965; Usher et al., 1969). For example, this last group of authors clearly showed that stress, after destruction of the septum, induces a much smaller rise in the corticosteroid level in the peripheral blood plasma than in intact animals. Meanwhile, no such difference is found if the function of the pituitary-adrenal system is assessed by the weight of the adrenals.

Another important source of the differences between the results is the time factor. For example, on the 3rd day after destruction of the anterior part of the median eminence the mean diameter of the nuclear mass in the zona fasciculate of the rat adrenal is lowered, whereas on the 7th day it becomes normal (Saba et al., 1962a). In rats with a focus of destruction in the hypothalamus the response was blocked 1 h after stress, but after 4 h the blood corticosterone level was almost the same as in animals with an intact brain (Brodish, 1964a). On the 3rd day after destruction of the anterior part of the median eminence the corticosterone secretion in rats was increased (Saba et al., 1962b,c), whereas after 7 days it fell considerably below the initial level (Saba et al., 1962b; Carnicelli et al., 1964; Tronchetti et al., 1965).

When experiments are performed, it must also be borne in mind that certain injuries of the brain can disturb the 24-hourly rhythm of fluctuations in the blood plasma corticosteroid level. The time of day at which the experiments are performed can therefore be another source of incorrect conclusions (Moberg et al., 1971; Seggie and Brown, 1971).

The operative techniques used in the experiments may also be responsible for differences between the results. To begin with, in lesions of the hypothalamus it is difficult to assess precisely what is damaged: the nervous centers, the conducting fibers, or both at once. In such cases it is difficult, therefore, to assess the specific role of the nervous structures of the hypothalamus in the regulation of pituitary-adrenal function. Another source of error may be the estimation of the size of the damaged area: the larger the injury, the more difficult it is to compare the effects of what were apparently identical injuries. After extensive destruction, the disturbances become more complex still and, in addition, as Szentagothai has pointed out, the investigator unconsciously estimates the difference between the results of two analogous lesions as proportional to their size (Szentagothai et al., 1965).

Electrical stimulation with the aid of permanently implanted electrodes can give information on the specific role of the nervous structures of the brain. However, if this method is used on animals with an intact brain, the possibility of an indirect effect of stimulation through the transmission of nervous impulses along descending pathways from the hypothalamus through the spinal cord to the periphery, followed by a secondary response affecting the pituitary-adrenal system through changes in the functional state of the organs or of metabolism, cannot be ruled out. This is shown by experiments in which electrical stimulation of different parts of the hypothalamus (Folkow and Euler, 1954; McHugh et al., 1966), of the cerebral cortex (Ferguson et al., 1957; Euler and Folkow, 1958) or cerebellum (Hirai, 1966) was accompanied by changes in the blood pressure and in the secretion of catecholamines from the adrenal medulla.

The actual insertion of electrodes, at least into certain areas of the hypothalamus (the preoptic region, suprachiasmatic nucleus, arcuate nucleus of the hypothalamus, median eminence) produces a statistically significant increase in the peripheral blood corticosterone level in rats (D'Angelo et al., 1964).

During electrical stimulation, the possibility cannot be ruled out that the ultimate response of the pituitary-adrenal system may depend on the functional "tuning" of the nervous structures at the time of stimulation.

The reader will be aware of criticism of the classification of stressors suggested by Fortier (Fortier, 1951a,b,c; Fortier et al., 1957) into systematic (induced by adrenalin, histamine, and so on) and psychic (induced, for example, by sound). In the opinion of Sayers et al. (1958) it is more correct to classify stressors by intensity than by quality. Contradictions between experimental results may be associated with the use of stressors of different intensity. For example, depending on the parameters of the electric current used, on which more will be said below, Endröczi and Lissak (1962) observed differences in the response of the pituitary-adrenal system after electrical stimulation of the hippocampus.

Finally, differences in the results of experiments may be due to the species of experimental animal or to the class to which it belongs.

The Hypothalamus as a Whole and as Part of the Central Mechanism of Regulation

The many facts described above indicate that functional changes in the pituitary-adrenal system arise in the presence of destruction or electrical stimulation of nearly all the nuclear formations of the hypothalamus, which lie in close juxtaposition in its ventromedial part. Disturbance of any discrete center of this zone causes changes in the pituitary-adrenocortical system, evidently in connection with disturbance of the anatomical and functional integrity of the hypothalamus and its wide range of connections with other parts of the brain. Consequently, the hypothalamus affects the pituitary-adrenal system as a single entity, and it is part of the general homeostatic mechanism of the central nervous system. The investigations described below confirm this point of view.

The medial zone of the hypothalamus in rats was conventionally divided by Brodish (1963) into four successive areas, each of which was destroyed by electrolysis. This worker observed a decrease in the intensity of the response of the pituitary-adrenal system to stress (ether anesthesia) 24 h after the operation. The corticosterone concentration in the peripheral blood was 55–80% of its level in animals with an intact hypothalamus. If the lesion covered two areas or more, the corticosterone concentration fell to 25–40% of the normal. On the basis of these results, Brodish (1963) concluded that the hypothalamus more probably contains diffusely arranged nerve centers or a nerve net than a localized, discrete nerve center, consisting of concentrations of neurons. The diffuse centers constitute the morphological basis of the mechanism regulating ACTH secretion. This diffuse system extends from the mammillary bodies to the optic chiasma in the ventromedial part of the hypothalamus (destruction of the lateral areas of the hypothalamus was ineffective), with the highest concentration of neurons in the region of the tuber cinereum.

Brodish (1964a,b) carried out some very interesting experiments which showed that lesions of different extent in the ventral hypothalamus of rats delay the rapid increase in ACTH secretion in response to stress. However, the blood corticosterone level in these animals rises, despite its delay, up to 90% of the concentration found in normal animals. Brodish points out that, notwithstanding the extensive lesions of the basal part of the hypothalamus (in some animals the zone of destruction extended from the optic chiasma to the mammillary bodies), the secretion of corticosterone was higher than in hypophysectomized rats. The latest experiments suggest the existence of an additional extra-hypothalamic mechanism participating in the regulation of function of the pituitary-adrenal system.

These experiments by Brodish (1963) agree with those (Moll, 1959, 1960) which showed that pinpoint lesions of the hypothalamus in the midline are accompanied by a decrease in weight of the adrenal and a tendency toward inhibition of the compensatory hypertrophy of the residual adrenal 6 days after removal of the opposite gland. In each animal the hypothalamus was injured at one point only and the extent of the damage did not exceed 1 mm in diameter. This very slight destruction of nerve tissue, in Moll's opinion, could not be the cause of inhibition of secretion of the corticotrophin-releasing factor, for the zone of the hypothalamus connected with its production is much larger than the zone occupied by a single lesion. Moll (1959, 1960) considers that the cause of the decrease in weight and the inadequacy of hypertrophy in the adrenal is not so much destruction of the neuroendocrine centers concerned with the secretion of CRF as a general disturbance of the nervous and neuroendocrine connections of the integral homeostatic mechanism of the brain stem.

There is no doubt that the hypothalamus, an extremely important autonomic center, is under the constant influence of extra-hypothalamic brain structures. The close relationships between the hypothalamus and other structures of the central nervous system are demonstrated, on the one hand, by its anatomical connections and, on the other hand, by many experiments which have shown the influence of different parts of the brain on the function of the hypothalamo-hypophyseo-adrenal system. For the reasons given above, the hypothalamus must be regarded as part of an integral mechanism of the central nervous system controlling the adrenocorticotrophic function of the pituitary gland. Although at the present time it is impossible to paint a clear picture of the situation as a whole, there is no doubt that the limbic system and mesencephalon participate in the regulation of the pituitary-adrenal complex.

The Limbic System of the Telencephalon

The olfactory brain for a long time was neglected by physiologists, and only in the last decade has it begun to receive its due attention by specialists of different kinds. The olfactory brain has attracted the attention of physiologists because, as we now know, many of its nervous mechanisms have nonolfactory functions (one of the reasons for neglect of the rhinencephalon for so long was that anatomists had a tendency to regard it purely as the anatomical site of the function of olfaction), for they are concerned with the formation of the emotional behavior of animals and man. Nowdays, therefore instead of being called the "olfactory brain," the structures of the rhinencephalon are called by the term "limbic system" introduced by McLean (1952).

The Amygdalar Complex

After electrical destruction of the amygdalar complex, an increase in pituitary-adrenal activity is observed. These results were first obtained in monkeys (Mason, 1958, 1959a) and then confirmed in cats, dogs, rats (Endröczi et al., 1959; Okinaka, 1961), and man (Mandell et al., 1963). In man a particularly marked increase in the concentration of 17-hydroxycorticosteroids in the blood and urine is observed after stimulation of the inferolateral part of the amygdala (Rubin et al., 1966).

If the structures of the amygdalar complex in monkeys are stimulated interruptedly for 90 min, the 17-hydroxycorticosteroid level in the peripheral blood plasma rises rapidly. The increased adrenocortical activity persists for several days (Mason et al., 1960). After electrical stimulation of the amygdalar complex of cats and dogs for 5-10 min, besides an increase in the concentration of hydrocortisone and corticosterone, large quantities of androgenic components are also found in the blood flowing from the adrenals. This type of secretion differs from the ordinary type when, after administration of ACTH, application of stressors, or stimulation of the mesencephalic reticular formation and posterior hypothalamus, the secretion of only those corticosteroids which are found in the blood under normal conditions is increased. These changes in corticosteroid secretion are regarded as specific adaptation of the adrenal cortex, through which the gland participates in the maintenance ofhomeostasis (Endröczi and Lissak, 1960).

The greatest stimulation of the function of the pituitary-adrenal system is found during electrical stimulation of those parts of the amygdalar complex which are concerned with the manifestation of emotions (Setekleiv et al., 1961). However, the increase in adrenocortical function is not secondary to the emotional responses, for in experiments on cats under dial anesthesia, which blocks the manifestation of emotions, after electrical stimulation of the amygdalar complex there was an increase in the blood corticosteroid level as before (Setekleiv et al., 1961). The same conclusion can be drawn from the results of experiments on unanesthetized monkeys (Mason, 1959a), in which under similar experimental conditions, despite a definite increase in the 17-hydroxycorticosteroid concentration in the peripheral blood plasma, neither emotional nor somatomotor disturbances are observed.

The amygdalar complex includes zones which stimulate and also zones which inhibit the pituitary-adrenal system. In an acute experiment on "encéphale isolé" cats, electrical stimulation of the medial part of the basal nuclei for 30 sec induced an increase in the secretion of corticosteriods into the blood flowing from the adrenal gland, whereas stimulation of the lateral nuclei or lateral part of the basal nuclei of the amygdalar complex inhibited secretion. Meanwhile, during stimulation of the zone at the boundary between the basolateral and basomedial nuclei, the blood corticosteriod concentration was unchanged (Slusher and Hyde, 1961b). Electrical stimulation of the medial part of the amygdalar complex is accompanied by rapid activation of the pituitary-adrenal system in dogs also (Redgate, 1970).

An important investigation of the detailed localization of nuclear groups related to the pituitary-adrenal system in the amygdala has recently been published (Matheson et al., 1971). This work was carried out on unanesthetized cats which could behave freely. The experiments showed that electrical stimulation of the corticomedial, basal, and lateral nuclei of the amygdalar complex is accompanied, first, by activation of the pituitary-adrenal system. An inhibitory effect occurred mainly in response to stimulation of the corticomedial nuclei. Meanwhile, stimulation of the anterior portion of the amygdala caused no change in the blood corticosteroid concentration. In this investigation an important fact was discovered: the direction of the changes in the blood corticosteroid concentration depended on the initial blood level of adrenal hormones before stimulation. If the initial level was low, electrical stimulation evoked an increase in the corticosteroid concentration, but if the initial concentration was high, the hormone level was unchanged or was lowered. These results provide new information on the role of the amygdala as a modulator of pituitary-adrenal activity.

In attempts to discover the role of this structure, several workers have studied the state of the adrenal cortex after amygdalectomy. The results obtained by this operation on different animals are highly contradictory and require clarification. In dogs and cats, for instance, bilateral removal of the amygdalar complex causes an increase in the secretion of hydrocortisone and corticosterone into the blood stream (Martin et al., 1958). Similar results have been obtained in mice (Eleftheriou et al., 1966).. Presumably in these animals the amygdalar complex contains chiefly elements which inhibit the function of the pituitary-adrenal system. Meanwhile, in rats amygdalectomy is not followed by an increase in the function of the pituitary-adrenal system, but against the background of stress induced by immobilization of these animals, the increase in the blood corticosterone level is delayed (Knigge, 1961).

In chronic experiments on rats after bilateral electrolytic destruction of the amygdalar complex, the adrenal glands at first increase in size up to a maximum on the 13th day after the operation, after which they decrease in weight (Yamada and Greer, 1960). These workers consider that either the response of the adrenals is nonspecific in character and is connected with the stress of the operation, or during the operation other parts of the central nervous system, which in animals with an intact brain have an inhibitory effect on ACTH secretion, are also damaged. They conclude that the amygdalar complex in rats does not play an essential role in the mechanisms of ACTH secretion. However, it is impossible to draw categorical conclusions regarding the state of the functions of the adrenal glands purely on the basis of a study of their weight.

These results obtained by different workers thus give the impression that the amygdalar complex contains dissimilar groups of neurons (excitation of some groups causes stimulation, while excitation of others causes inhibition of the function of the pituitary-adrenal complex), and that activating influences are predominant.

Meanwhile analogous structures, whose excitation is accompanied mainly by an inhibitory effect on the adrenal cortex, are found in the limbic system. This statement applies primarily to the hippocampus.

The Hippocampus

Evidence of the inhibitory influence of the hippocampus on pituitary-adrenal function is given by the results of experiments on monkeys, dogs, cats, rabbits, and rats (Mason, 1958; Endröczi et al., 1959; Endröczi and Lissak, 1960; Okinaka et al., 1960a,b).

Mason (1958) showed that stimulation of the medial region of the anterior part of the hippocampus and the entorrhinal cortex is not accompanied by any increase in the 17-hydroxycorticosteroid concentration in the peripheral blood plasma of monkeys, while after 24-48 h their level is lowered to such a degree that it does not even reach the lower limit of the corticosteroid concentration in control animals. No increase in the secretion of corticosteroids was found in the adrenal vein of "encéphale isolé" cats by other workers 15-60 min after stimulation of the entorrhinal cortex (Slusher and Hyde, 1961b). Meanwhile stimulation of the uncus hippocampi caused inhibition of adrenocortical function (Slusher and Hyde, 1961b). Okinaka et al. (1960a,b) found that the ACTH level in the jugular vein of dogs falls after hippocampal stimulation; meanwhile the secretion of corticosteroids into the blood flowing from the adrenal was changed by a lesser degree.

Electrical stimulation of the hippocampus depresses the function of the pituitary-adrenal system not only in a state of rest, but also during stress, as Porter (1954) showed for the first time. It was later discovered that, despite exposure to stress factors (electrical stimulation of the animal's forelimb, injection of adrenalin, histamine, or formalin), which were applied at the beginning of a 15-min period of stimulation of the dorsal part of the hippocampus, inhibition of the response of the pituitary-adrenocortical system was observed. This was reflected in a decrease in the rate of secretion of corticosteroids into the venous blood flowing from the adrenal gland in cats, a decrease in the ascorbic acid content in the adrenals of rats, and a lymphocytopenic reaction in rabbits (Endröczi et al., 1959; Lissak and Endröczi, 1965).

After stimulation of the hippocampus, adrenocortical function is altered in man also. In patients with psychomotor epilepsy, stimulation of the hippocampus through temporarily implanted electrodes causes a decrease in the corticosteroid concentration in the peripheral blood plasma and in the urine (Mandell et al., 1963; Rubin et al., 1966).

In the analysis of the results obtained after electrical stimulation of the anatomical structures of the brain, allowance must always be made for the parameters of the current used, for the response of the pituitary-adrenal system is particularly dependent on them. For example, stimulation of the dorsal part of the hippocampus by electric pulses with a frequency of 12-36/sec causes inhibition of secretion of adrenocortical hormones in response to nociceptive stimulation. Stimulation of the same area by pulses with a frequency of 120/sec has a weak effect on stress due to nociceptive stimulation, while stimulation at a frequency of 240/sec is accompanied by increased function of the pituitary-adrenal complex (Endröczi and Lissak, 1962). Activation of the adrenal cortex after electrical stimulation of the posterior part of the hippocampus at a frequency of 40-100/sec has also been found in monkeys (Smith and Root, 1971).

Experiments such as these suggest that the influence of this part of the limbic system is not always reflected in an inhibitory effect on the pituitary-adrenal system. Clearly the intensity of excitation of the neural elements of the hippocampus plays by no means a minor role, for it determines the character of the effect. Conclusions obtained from experiments involving electrical stimulation must evidently, therefore, be confirmed by other methods.

Whereas stimulation of the region of the cornu ammonis leads to inhibition of pituitary-adrenal function, its destruction is followed by the opposite effect. In cats, for instance, bilateral destruction of the area of the hippocampus forming the side of the lateral ventricle induces a marked increase in corticosteroid secretion into the adrenal vein by comparison with the initial level (Fendler et al., 1961). Similar results have also been obtained in rats, when after bilateral destruction of the hippocampus there is a further increase in the blood corticosterone concentration in response to stress induced by immobilization for 30 min (Knigge, 1961). After destruction of the dorsal part of the hippocampus, stress is accompanied by a greater fall in the ascorbic acid concentration in the adrenals of rats than in control animals or in rats with destruction of the cerebral cortex above the hippocampus (Kim and Kim, 1961). In the case of chronic stress, when the ascorbic acid concentration in the adrenals usually rises (Long, 1947), after injury to the hippocampus the ascorbic acid concentration increased by a smaller amount than in control rats, especially those in which only the cerebral cortex was damaged.

Since the majority of afferent fibers reach the hypothalamus from the hippocampus through the fornix, division of the fornix is regarded as equivalent to removal of the hippocampus. However, this operation in rats does not alter the initial blood corticosterone level and does not block the response of the pituitary-adrenal system to administration of reserpine or to cold stress. In rats the primary pathway from the hippocampus via the fornix is regarded as not essential for manifestation of the reflex to the pituitary (Revzin and Maickell, 1966).

The Septum

A change in adrenocortical function is observed after injury to the septum, although in these experiments the increase in the blood corticosteroid level does not reach values such as are found after injury, for example, to the gyrus cinguli (Bohus, 1961). An increase in the intensity of the secretory function of the adrenal cortex is observed after injury to the anterior portions of the septum. If, however, the posterior portion is damaged, the corticosteroid secretion falls (Endröczi and Lissak, 1960), while the response to stress is diminished or blocked (Ahren, 1962). Weakening or absence of the response to stress, as judged by the corticosteroid level in the peripheral blood plasma, is also observed after destruction of the lateral portion of the septum. Meanwhile, against the background of stress the ACTH concentration in the pituitary gland of these animals is increased (Usher et al., 1967, 1969). Electrical stimulation of the septum causes inhibition of corticosteroid secretion into the adrenal vein of "encéphale isolé" cats (Slusher and Hyde, 1961b).

The septum evidently contains neural elements whose stimulation produces different effects on the functions of the pituitary-adrenal complex. For example, in cats electrical stimulation of the ventral portion is accompanied by inhibition of corticosteroid secretion, while stimulation of the dorsal portion or descending part of the fornix either does not cause any significant inhibition or the rate of corticosteroid secretion is increased (Endröczi and Lissak, 1963). In dogs, after electrical stimulation of the dorsal portions of the septum, the ACTH level in the peripheral blood plasma rises even during the first 1-4 min (Redgate, 1970). In monkeys in self-stimulation experiments using electrodes implanted into the region of the septum, an increase in the 17-hydroxycorticosteroid concentration in the urine is also observed (McHugh et al., 1966). According to Katsuki (1961), electrical stimulation of the septal region in rabbits is accompanied by an increase, whereas its destruction is followed by a decrease, in weight of the adrenals. A decrease in weight of the adrenal glands after destruction of the septum has also been found in rats (Moll, 1959), although not by all investigators (Montgomery and Berkut, 1969; Usher et al., 1969).

Other Structures of the Limbic System

Information is also given in the literature on the effects of other parts of the limbic system than those mentioned above on activity of the pituitary-adrenocortical system. In response to stimulation of the orbital surface of the frontal lobe by permanently implanted electrodes in dogs, monkeys, and rats, Csordas et al. (1952), Porter (1954), Endröczi et al. (1958b), and Okinaka (Okinaka et al., 1960a,b; Okinaka, 1961) found evidence of stimulation of adrenocortical function. It is interesting to note that electrical stimulation of this region of the brain causes an increase in corticosteroid secretion into the blood flowing from the adrenals without affecting the secretion of catecholamines (Goldfien and Ganong, 1962). Increased function of the pituitary-adrenal complex induced by stimulation of the orbital surface of the frontal lobes is thus not due to an increase in the secretion of adrenalin from the adrenal medulla. This is confirmed by experiments in which division of the splanchnic nerves or demedullation of the adrenals caused no change in the response of this system to electrical stimulation by comparison with the effect in normal animals (Lissak and Endröczi, 1965).

Excision or destruction of the orbital part of the frontal lobes is followed by inhibition of the response of the pituitary-adrenal system to stress (Endröczi and Nagy, 1951; Ahren, 1962).

Evidence for the influence of the gyrus cinguli on adrenocortical function is scanty and contradictory. It has been shown that electrical stimulation of the cortex of the anterior part of the gyrus cinguili increases the 17-hydroxycorticosteroid concentration in the peripheral blood of cats on the average by 77.7%. Meanwhile, similar stimulation of the middle and posterior parts of this gyrus does not change the blood hormone level (Setekleiv et al., 1961). Stimulation of the anterior and posterior parts of the gyrus cinguli in dogs under morphine anesthesia rapidly increases the blood ACTH concentration (Ibayashi et al., 1963). Finally, destruction of the anterodorsal part of this gyrus in rats under chronic experimental conditions induces a marked increase in corticosteroid secretion (Bohus, 1961).

It thus now seems firmly established that many structures of the limbic system may have either a stimulating or an inhibitory effect of the pituitary-adrenal complex. The limbic system evidently participates in the mechanisms controlling pituitary-adrenal activity.

The Role of the Mesencephalon

The role of the mesencephalon in the regulation of pituitary-adrenal function has been studied to a much lesser degree than that of the hypothalamus. The large number of nuclei and of afferent and efferent fibers makes analysis of the results difficult.

The most general ideas of the role of the mesencephalon can be obtained after its experimental division. However, the facts obtained by these models are highly contradictory.

According to some workers (Tronchetti et al., 1964), a few hours after mesencephalic section the secretion of corticosterone in rats and its level in the peripheral blood are lowered. From 5 to 25 days after the operation the ascorbic acid concentration in the adrenals of rats is the same as in animals with an intact brain (Giuliani et al., 1961). In dogs, however, after mesencephalic section, as the work of Franschini et al. (1964) and Kendall et al. (1965) has shown, the corticosterone level rises. A similar increase is observed in dogs 24-48 h after section (Egdahl, 1962). Meanwhile, after the corresponding operation on cats the rate of aldosterone and corticosterone secretion differs only very slightly from the initial level determined in animals with an intact brain (Newman et al., 1958).

After mesencephalic section the pituitary-adrenal system

reacts to stress, although it does so differently to different stimuli. This indicates that the operation does not block its function (Giuliani et al., 1961).

Mesencephalic section experiments naturally do not answer the question whether the change in function of the pituitary-adrenal system is the result of disturbances in the structures of the mesencephalon or the result of more general disturbances arising after division of nearly all the sensory fibers, the pathways of the ascending reticular activating system, and efferent fibers. Nevertheless, the fact that experiments yield different results, combined with the anatomical and functional complexity of the mesencephalon, suggests that neural elements capable of stimulating as well as inhibiting the function of the pituitary-adrenal system may exist in the mesencephalon. This possibility has been confirmed experimentally.

The Tegmentum Mesencephali

After injury to the dorsal and ventral parts of the tegmentum immediately rostral to the pons, during the first 15 min after laparotomy there is an increase in the rate of corticosterone secretion into the venous blood flowing from the rat adrenal gland; from 15 to 30 min after stress in animals with injury to the dorsal tegmentum the secretion of corticosterone continues to rise, whereas in rats with disturbances in the region of the ventral tegmentum the rate falls (Slusher, 1960). When, under similar conditions, parts of the mesencephalon located more rostrally are damaged, the corticosterone secretion is unchanged. After destruction of the most anterior portions of the mesencephalon and caudal areas of the diencephalon, on the other hand, the response of the adrenal cortex to acute stress is inhibited (Slusher and Critchlow, 1959).

The rate of secretion of the corticosteroids also falls in cats 4 h after destruction of the ventral tegmentum mesencephali (Neuman et al., 1958). On the other hand, injury to the region of the mesencephalon around the aqueduct, involving the habenula and the stria medullaris, and sometimes the pineal body, causes an increase in corticosteroid secretion 4 h after the operation (Taylor and Farrell, 1962).

The results described above are in agreement with those obtained after electrical stimulation of the mesencephalon by permanently implanted electrodes: stimulation of the ventral tegmentum causes an increase in the rate of secretion of corticosteroids, while stimulation of the dorsal tegmentum causes a decrease (Endröczi and Lissak, 1963). These experiments suggest that the region of the ventral tegmentum contains neural elements principally stimulating the function of the pituitary-adrenal system, while most of those in the dorsal tegmentum are inhibitory in their effect. However, this hypothesis has been contradicted by the later work of Slusher (Slusher and Hyde, 1961a, 1966).

In acute experiments on "encéphale isolé" preparations, stimulation of the ventral tegmentum mesencephali for 30 sec caused a rapid fall in the rate of secretion of adrenocortical hormones into the blood stream (Slusher and Hyde, 1961a). In chronic experiments on cats with an intact brain, electrical stimulation of the ventral portion of the mesencephalon inhibited ACTH secretion, whereas stimulation of its dorsal part immediately below the superior colliculi increased the relief of ACTH into the blood stream (Slusher and Hyde, 1966). In dogs, electrical stimulation of the ventral tegmentum was accompanied by an increase in the blood level of ACTH in only one of the four experiments (Redgate, 1970). Further experiments are necessary to account for the differences between these results.

The Mesencephalic Reticular Formation

The mesencephalic reticular formation plays an important role in the regulation of activity of the pituitary-adrenal system. Its stimulation at the level of the superior colliculi in cats increases the rate of secretion of corticosteroids into the blood flowing from the adrenal (Endröczi and Lissak, 1960). Similar results have been obtained in dogs (Okinaka et al., 1960b), in which an increase in the ACTH level in the peripheral blood plasma is also found (Redgate, 1970). The opposite effect is found after destruction of the reticular formation: 4 h after the operation in cats, inhibition of the secretion of adrenocortical hormones is found (Newman et al., 1958). Rats with injuries in the region of the ascending activating reticular system do not respond to stress produced by immobilization by an increase in the function of the pituitary-adrenal complex (Knigge and Hays, 1963).

There is thus no doubt about the importance of the mesencephalon in the regulation of the pituitary-adrenal system under normal conditions and in response to stress. The mesencephalon contains neural elements capable of stimulating and inhibiting this system. These centers are possibly distributed diffusely in the mesencephalon (Slusher and Hyde, 1966). However, the question of the actual localization of the various structures requires further study.

Emotional Factors and the Pituitary-Adrenocortical System

The dominant role of the hypothalamus as well as the great importance of the limbic structures of the telencephalon and mesencephalon in the control of the adrenocorticotrophic function of the pituitary, on the one hand, and the direct participation of these parts of the brain in the formation of the emotional behavior of man and animals, on the other hand, suggest that emotional factors and the pituitary-adrenocortical system must be closely interconnected. The influence of the emotions, i.e., specific forms of changes in behavior, on the pituitary-adrenal complex has been studied in observations of man and experiments on animals, and interesting results have been obtained.

A significant difference has been found in the 17-hydroxycorticosteroid levels of persons with different degrees of excitability: the mean corticosteroid level in healthy inexcitable subjects is 9.2 μg%, compared with 13 μg% in easily excitable persons (Fiorica and Muchl, 1962). Psychiatric patients troubled by a feeling of expectation of something unpleasant or by restlessness (Mason, 1959b) or melancholia (Gibbons and McHugh, 1962) have an abnormally high level of corticosteroids in the blood and urine. Meanwhile, in persons in a tranquil and peaceful frame of mind the hormone level in the blood and urine is low (Handlon et al., 1962).

The corticosteroid level found in human subjects and, evidently, in animals in a state of tranquility is evidently elevated to some degree by emotional environmental factors. A demonstration of stylized films lowers the 17-hydroxycorticosteroid level in the blood. It is suggested that this decrease may be connected with the concentration of attention on "innocuous" stimuli which distract the subject from possible emotional, stress-inducing stimuli of endogenous nature (Handlon et al., 1962). The lowering of the hydrocortisone concentration in healthy persons in a hypnotic state is evidently similar in its origin (Persky et al., 1959). On the other hand, anxiety induced by suggestion in a state of hypnosis leads to an increase in the blood corticosterone concentration (Levitt et al., 1964). Emotions of a negative character (anxiety, fear, horror, etc.) as a rule stimulate adrenocortical function. In healthy subjects the blood corticosteroid concentration rises after watching "horror" films in the course of which 30 murders were shown in 1 hour. Under these conditions, the more closely dependent the response of the endocrine glands on the subject's mental activity, the more strongly his emotional state is expressed during the showing of the film (Euler et al., 1959). Simulation of the possibility of an accident during an unexpected explosion evokes an increase in the excretion of 17-hydroxycorticosteroids in the urine (Berkun, 1964). A marked increase in the intensity of adrenocortical function is observed in soldiers on the battlefield (Pace et al., 1956). Anticipation of an impending surgical operation has the same effect (Franksson and Gemzell, 1955).

It is claimed that the secretion of ACTH is determined, not by the level of motor activity, but by its "emotional charge" by its affective component (Endröczi and Lissak, 1960). On the day of the race, adrenocortical function of athletes is at a higher intensity than in the control period (before training) or on days of simple training exercises (Hill et al., 1955). The blood corticosteroid level is raised in the competitors during boat races. This increase must be due to psychological agitation rather than to physical exertion because it was present throughout the period of the race and was not found on ordinary training days, when the intensity of physical exertion was the same. Furthermore, the elevation of the corticosteroid level was less marked in the winning crew (Hill et al., 1956).

A close connection between emotional responses and the pituitary-adrenal system has also been demonstrated in experiments on animals. Stress induced by prolonged exposure to a high-pitched sound is accompanied by an increase in the blood corticosterone concentration in rats (Henkin and Knigge, 1963). Similar results have been obtained after stress induced by noise or vibration (Kokka et al., 1972).

If untrained monkeys are secured on a laboratory bench, their blood 17-hydroxycorticosteroid level rises considerably. On Saturdays and Sundays, however, when stimulating factors in the laboratory are much less evident, the concentration of corticosteroids in the blood and urine falls (Mason et al., 1957a,b). Evidence of the effect of mental stress on the function of the pituitary-adrenal complex in primates has also been obtained by more recent investigations (Mason, 1968; Brown et al., 1971a,b).

There is considerable documented evidence to show that interaction between animals in groups has a considerable influence on the function of the pituitary-adrenal system (see Christian, 1950, 1963; Christian et al., 1965; Welch, 1965). Investigations of interpopulation relationships have shown that any population or group of animals is not homogeneous, but its members are in complex behavioral interaction with each other. Hierarchical relations are most clearly visible in birds (Etkin, 1967). In mammals the hierarchical structure of a community of animals is often manifested in the form of very complex relations between its individual members, and it may be difficult to detect. However, the existence of dominant and subordinate individuals is now regarded as an undoubted fact, and it is evidently one of the basic features of life of any community of mammals.

Investigations have shown a reciprocal relationship between the state of the pituitary-adrenal system and the animal's position in the community. For example, if mice normally kept in isolation were put together in groups of 6 animals for 4 hours daily over a period of 10 days, there was an increase in the weight of the adrenals, and the lower the social rank of the animal the greater the increase which was observed (Davis and Christian, 1957). These results have recently been confirmed by experiments in which more objective criteria were used to assess hierarchical behavior and the functions of the pituitary-adrenal system (Naumenko et al., 1971; Popova, 1971; Popova and Naumenko, 1972).

First, just as in the experiments of other workers (Christian, 1950, 1963; Mason, 1959b; Christian et al., 1965), it was found that the level of functional activity of the pituitary-adrenal system is higher in animals in a group than in isolation. The fact that the corticosterone level in animals in a group was raised because of contact between the individuals in the group is confirmed by experiments in which a group of rats was decapitated at the usual time, but in an inactive drowsy state (Popova, 1971). Under these conditions the mean corticosterone level was found to be much lower than in corresponding animals in a waking state, but it did

TABLE 1. Corticosterone Concentration in Peripheral Blood Plasma of Male Albino Rats in States of Sleep and Waking

	Mean corticosterone level (M ± m)			
	Sleep		Waking	
In isolation	3.6± 0.3 (10)*	P > 0.05	5.7 ± 1.0 (4)	p < 0.05
In a group	5.7± 1.4 (10)		15.9 ± 4.2 (10)	

*Number of estimations.

not differ significantly from the hormone concentration in isolated rats (Table 1).

Observations on the natural behavior of male albino rats kept in groups of 10 in a cage and comparison of the group activity with the blood corticosteroid level showed no clear linear relationship between these indices. However, comparison of two extreme groups – those with the highest and those with the lowest indices of dominance – showed that the corticosterone level in the first group was only half that in the second (Popova, 1971; Popova and Naumenko, 1972) (Table 2).

It is important to note that the increased activity of the pituitary-adrenal system in the subordinate animals was due primarily to emotional and not to physical factors. There is much evidence in support of this conclusion. If male mice of line C57BL/10y

Table 2. Group Relationships and Corticosterone Concentration in Peripheral Blood Plasma of Male Albino Rats

Index of dominance* M ± m	Index of social activity†	Corticosterone level, μg%	Number of animals
2.2 ± 0.36	121.0 ± 40.0	6.9 ± 0.91	15
0.4 ± 0.05	14.2 ± 3.2	13.0 ± 1.73	17
P < 0.001	P < 0.02	P < 0.01	

*Ratio between total number of attacks by a particular rat and number of attacks made on it.

†Index of dominance multiplied by total number of fights involving the particular rat.

are placed daily in the same cage with a trained male "fighter" of the same line, the number of fights is reduced, whereas the blood corticosterone level and the ACTH concentration in the pituitary, determined immediately after removal of the subordinate animal from the cell with the "fighter," remains high (Bronson and Eleftheriou, 1965a). In a similar situation, merely the sight of the aggressive animal is sufficient to stimulate adrenocortical function. If a mouse which had already experienced being placed together with the "fighter" mouse was again placed in the same case with it, the corticosterone level in its peripheral blood increased despite the fact that the "fighter" and the stranger were separated by a wire screen (Bronson and Eleftheriou, 1965b).

Counting the number of scratches and scars on the mice from the group showed no correlation between these indices and the weight of the adrenals (Christian, 1959). Hypertrophy of the adrenals was found in animals taken from communities in which there were no fights. In other communities, dominant brown rats took part in fights more often than subordinate animals, although hypertrophy of the adrenals was found in the latter (Barnett, 1963).

Consequently, the facts described above would seem to prove conclusively that certain emotional states are accompanied by activation of the pituitary-adrenal complex.

The change in the state of the pituitary-adrenal system depends on to what extent the animal is adapted to the unusual stimuli. For example, low-pitched noise from an airplane engine in motion caused an increase in adrenocortical function only in subjects who were not accustomed to this noise (Hale, 1952). However, the response of the pituitary-adrenal system to a new situation is quickly extinguished. A marked and prolonged increase in the circulating blood corticosterone level is found in mice when attacked for the first time by a specially trained "fighter" mouse. Adaptation to this situation takes place very quickly, and whereas after the first attack the increased corticosterone level persists for 24 h, after the fourth attack signs of adrenocortical stimulation can be seen for not more than 3 h (Bronson and Eleftheriou, 1965c).

An increase in adrenocortical function can also be obtained by a conditioned-reflex mechanism, as has been shown in monkeys. A loud noise, combined with electric shocks, after several repetitions itself began to evoke the maximal increase in the blood 17-

hydroxycorticosteroid level in the animals (Mason et al., 1957a). There is a definite relationship between the resting blood corticosteroid concentration and the degree of conditioning. In cats with clearly defined conditioned-reflex activity, the rate of secretion of corticosteroids into the blood stream from the adrenal is low, but in animals with low conditioned-reflex activity the rate of secretion is higher (Lissak and Endröczi, 1964).

It was stated previously that the response of the pituitary-adrenal system to a new situation is rapidly extinguished. However, in conditioned-reflex experiments this response persists for many months. It is assumed that the mechanism of hypothalamic activation of ACTH is different in the two cases (Gellhorn and Loofbourrow, 1963).

It is thus firmly established that many of the levels of the central nervous system from the hypothalamus to the cerebral cortex are concerned with the control of activity of the hypothalamo-hypophyseo-adrenal system. Although it is difficult at present to draw up a general scheme for the operation of the complete brain mechanism, it is evident that wherever the impulses arise they must eventually reach the hypophyseal zone of the hypothalamus, which secretes corticotrophin-releasing factor.

The identity of the mediators transmitting the effect of nervous stimuli arising in the brain in response to the action of stressors of extremely diverse origin is a question to which the answer must be found. To put it another way, what is the role of the chemically reactive structures of the brain in the regulation of the hypothalamo-hypophyseo-adrenal complex? Are the adrenergic, cholinergic, and serotoninergic structures concerned in the regulation of this system or only some of them, and in what ways do they act: do the chemically reactive structures have a direct central influence on the hypophyseal zone of the hypothalamus or do they have an indirect influence by descending neurons acting through the spinal cord and corresponding peripheral mechanisms? In the next chapters I shall try to answer these questions as far as possible.

Chapter II

Adrenergic Structures and Regulation of Pituitary-Adrenal Function

The role of mediators in the transmission of the nervous impulse from the preganglionic fiber to the ganglion and from the postganglionic fiber to the effector organ is no longer in question. The principal mediators for peripheral nerves – acetylcholine and noradrenalin – are also sufficiently well known. The problem of the transmission of nervous impulses and the types of connection between single cells in the central nervous system is a much more complex matter. The role of transmitters in this case has been ascribed to a number of different chemical agents, but at the present time three biologically active substances are receiving most attention and are regarded as the most probable mediators. These are, first, compounds whose mediator role for peripheral organs is already established, viz., acetylcholine and noradrenalin, and in addition, the biogenic amine serotonin.* These compounds are endowed to the greatest possible degree with the properties essential to substances claiming to act as mediators in the central nervous system:

*Recent investigations have drawn attention to the possible mediator role of dopamine, the precursor in the biological synthesis of noradrenalin (McGeer, 1963; Rossum, 1963; Sourkes, 1964; Dahlström and Fuxe, 1965a; Hillarp et al., 1966a,b; Bertler and Rosengren, 1966; Hornykiewicz, 1966; King and Thomas, 1968). Particularly large quantities of dopamine are found in the region of the median eminence of the hypothalamus (Rinne and Sonninen, 1968). Recent work has demonstrated the central effects of this biogenic amine on the hypothalamo-hypophyseo-gonadal system (Fuxe et al., 1970a; Kordon and Glowinski, 1970). However, the possible role of dopamine in the regulation of the hypothalamo-hypophyseo-adrenal system is very debatable (Fuxe et al., 1970b).

1. All three biologically active substances – acetylcholine, noradrenalin, and serotonin – are present in neurons of the brain (Hebb and Whittaker, 1958; Glowinski and Iversen, 1966a,b; Hillarp et al., 1966a,b; Whittaker, 1966; Björklund et al., 1971). It is a very significant fact that these monoamines and acetylcholine are located principally in synaptic endings (Dahlström and Fuxe, 1965a; De Robertis et al., 1965; Aghajanian and Bloom, 1966, 1967; Hillarp et al., 1966a,b; Whittaker, 1966; Anden et al., 1969; Clementi et al., 1970).

2. Enzyme systems for the synthesis and inactivation of acetylcholine, the catecholamines, and serotonin are present in the central nervous system (Bogdanski and Udenfriend, 1965; Udenfriend et al., 1957b; Axelrod et al., 1959; Weiner, 1960, 1964; Brodie and Beaven, 1963; Sourkes, 1964; Laverty and Sharman, 1965; Iversen and Glowinski, 1966).

3. Specific mechanisms for absorption and deposition of monoamines exist in the brain (Brodie and Beaven, 1963; Anden et al., 1969).

4. Microelectrophoretic investigations of the central nervous system have shown the presence of specific individual nerve cells responding to local application of acetylcholine, noradrenalin, and serotonin and, consequently, containing cholinergic, adrenergic, or serotoninergic structures (Bloom et al., 1963; Salmoiraghi et al., 1965; Salmoiraghi and Stefans, 1965).

Furthermore, recent investigations have brought us fairly close to an understanding of the nature of these structures or, as they are usually called, synaptic receptors (Woolley and Gommi, 1966; De Robertis, 1971; Pidevich, 1971; Wesemann et al., 1971). Interesting results have been obtained by De Robertis (1971), who showed that the synapses of the cat brain, the electric organs of certain fishes, and the neuromuscular synapses of the rat diaphragm contain a hydrophobic proteolipid with the properties of a receptor, described as a receptor proteolipid. Electron-microscopic investigations have shown that this receptor consists of rod-shaped macromolecules, which are grouped into bundles as a result of the action of acetylcholine. However, the receptor proteolipid is probably only part of the synaptic receptor, for it has low specificity. According to De Robertis, the receptor proteolipid responds to acetylcholine, cholinomimetics, cholinolytics, serotonin, histamine, and α- and β-adrenergic blocking drugs. Meanwhile evidence has also been obtained that sialic acid, in the form of sialoglycoproteins or sialoglycolipids, is a component of the serotonin receptors in the brain (Wesemann et al., 1971).

5. In response to the action of substances altering the functional state of the central nervous system, changes take place also in the concentration and activity of acetylcholine, catecholamines, and serotonin (Pletscher et al., 1956; Brodie, 1958; Brodie et al., 1959; Spector et al., 1960; Pletscher and Gey, 1961; Potter and Axelrod, 1963).

Noradrenalin as a Hypothetical Mediator in the Brain

Many workers have studied the distribution of catecholamines in the central nervous system; they have discovered a high content of these substances (especially noradrenalin) in the hypothalamus, the gray matter around the aqueduct of Sylvius, in the tegmentum mesencephali, the medulla, and the medial part of the thalamus. Details of the distribution and metabolism of catecholamines in the brain have been obtained not only by biochemical, but also by histochemical (Dahlström and Fuxe, 1965a,b; Dahlström, 1967), electron microscopic (De Robertis et al., 1965), microelectrophoretic (Salmoiraghi and Stefans, 1965), and isotopic (Glowinski and Iversen, 1966a,b; Iversen and Glowinskii, 1966; Aghajanian and Bloom, 1966; Carr and Moore, 1969; Persson and Waldeck, 1969) methods of investigation.

The main mass of the cell bodies of the monoaminergic neurons lies essentially in the lower part of the brain stem: the medula, pons, and mesencephalon (this applies not only to noradrenalin, but also to dopamine and serotonin). The nerve endings of these monoaminergic cells are found, however, in nearly all parts of the brain (Anden et al., 1969). Characteristically the highest noradrenalin concentrations are found in distinctive varices located near endings of the nerve fibers. Concentrations of noradrenalin can be seen in these varices in the form of granules or intraneuronal organelles, which are evidently depots of the amine.

The discovery of high concentrations of noradrenalin in the regions of central representation of the sympathetic nervous system led to the hope that this was more than pure coincidence.

Some particularly hopeful experiments were carried out by De Robertis et al. (1965) who isolated synaptic vesicles from the hypothalamus belonging to axons and synaptic complexes and which contained high concentrations of noradrenalin. On the basis of this distribution of noradrenalin, he considered that it acts as a true mediator.

The concentration of adrenalin in the brain is much lower, approximately 5-10% of the noradrenalin concentration, although the character of its distribution agrees closely with that of noradrenalin (Weiner, 1964).

Experiments in which noradrenalin was injected by electrophoresis showed a wide distribution of neurons sensitive to noradrenalin, although the ratio between the number of neurons sensitive and insensitive to it varies in different parts of the central nervous system. It was also shown that the same catecholamine may have opposite action on different adrenergic neurons depending on the type of the neuron studied, although an inhibitory effect is more usual (Salmoiraghi, 1966).

Hillarp et al. (1966a) sum up the results of their many investigations into the role of monoamines as mediators of nervous impulses in the brain by the conclusion that the basic characteristics and properties of peripheral and central adrenergic neurons are the same:

1. The presence of very typical synaptic endings containing virtually the whole reserves of the amines. Their content in other parts of the neurons is low.
2. Analogous effects follow division of a peripheral axon or of an axon of the central nervous system.
3. Central adrenergic neurons contain enzyme systems for the synthesis and breakdown of catecholamines: a) dopadecarboxylase, b) dopamine-β-oxidase, c) monoamine oxidase, essential for inactivation of intraneuronal amines.
4. They contain specific mechanisms responsible for the reception and accumulation of amines, including:
   a) Mechanisms controlling reception and concentration, found in the cell membranes of the intact neuron and characterized by extremely high efficiency and by insensitivity to reserpine although they are blocked by substances such as cocaine and desmethylimipramine.

   b) Mechanisms controlling the reception and storage of reserves in specific granules; they form the amine granules in the cell bodies and transport them to the nerve endings, and they are specifically blocked by reserpine.
5. They react to substances which modify catecholamine metabolism in neurons (reserpine, tetrabenazine, m-tyrosine, α-MMT, α-methyldopa, monoamine oxidase inhibitors).
6. Under the influence of the nervous impulse, amines may be secreted from the synaptic ending and their synthesis in it increased:
   a) This has been demonstrated biochemically in experiments *in vitro* in which electrical stimulation of the descending bulbospinal noradrenalin-containing neurons was accompanied by an increase in noradrenalin synthesis in the spinal cord and its increased elimination into the incubation medium;
   b) Histochemical experiments have shown that vesicles of nerve endings are presynaptic structures specialized for the synthesis, storage, and elimination of catecholamines.

However, the direct liberation of noradrenalin during neuronal activity has not yet been demonstrated in the brain as it has for peripheral nerves, although there is indirect evidence in support of this liberation (Glowinski and Baldessarini, 1966; Anden et al., 1969).

The strikingly long latent period and the slow response of central neurons to noradrenalin is puzzling (Salmoiraghi, 1966). Such a slow response would be expected if the compound produced its effect indirectly. It is interesting from this point of view to compare the responses of individual neurons to noradrenalin with their response to a relatively nonspecific stimulus: L-glutamic acid. The latent period after application of L-glutamic acid through the same electrodes as noradrenalin was in every case shorter than for noradrenalin itself (Crawford and Curtis, 1964). These observations have led to the view that the chief role of catecholamines in the central nervous system is that of neurohormones or modulators creating the background for the activity of the true mediator.

Effect of Catecholamines and Substances Exciting and Blocking Adrenergic Structures on the Pituitary-Adrenocortical System

Of all the possible regulators of activity of the pituitary-adrenal system there is no substance whose mediator role has been proved or postulated which has been studied so intensively, yet with such comparatively little success, as the catecholamines. The reason for the tremendous number of investigations into the effects of adrenalin and noradrenalin from the standpoint of the subject now under discussion is that the sympathetic nervous system has long been regarded as a participant in responses to stress (Ramey and Goldstein, 1957), and adrenalin was regarded by many workers, especially in the late 1940s and 1950s, as the principal stimulator of the pituitary-adrenal complex.

Catecholamines and Adrenomimetics

After the discovery of the stimulant effect of adrenalin on ACTH secretion in the 1940s (Vogt, 1944; Long and Fry, 1945), numerous investigations showed that administration of adrenalin and, to a lesser degree, noradrenalin led: 1) to a decrease in the number of eosinophils (Godowski, 1948; Hortling and Pekkarinen, 1949; McDermott et al., 1950b; Samuels, 1951; Bliss et al., 1951; Speirs, 1953) and lymphocytes (Malmejak et al., 1947; Gellhorn and Frank, 1948, 1949; Endröczi and Nagy, 1951; Schweizer, 1953; Eskin and Mikhailova, 1960) in the peripheral blood; 2) to a decrease in the concentration of cholesterol, lipids (Symington, 1951), and ascorbic acid (Long, 1947; Ronzoni and Reichlin, 1950; Brodish and Long, 1962) in the adrenals; 3) to an increase in the survival time of adrenalectomized rats after injection of blood taken from animals into which adrenalin was injected (Vogt, 1944, 1947; Pickford and Vogt, 1951); and 4) to an increase in the oxygen consumption and the intensity of aerobic glycolysis in the anterior lobe of the pituitary gland (Roberts and Keller, 1955).

The stimulant action of adrenalin on the pituitary-adrenal system has been demonstrated in rats (Long and Fry, 1945; Vogt, 1947; Danford and Danford, 1951; Symington, 1951), mice (Speirs and Meyer, 1949), dogs (Pickford and Vogt, 1951), guinea pigs (Burstein, 1952), cats (Porter, 1953a,b), sheep (Zarrow et al., 1952), and man (August and Gubner, 1949; Samuels, 1951; Bisso,

1953; Wolfson, 1953; Vernikos-Danellis and Marks, 1962). This effect has been obtained after intravenous (Vogt, 1947; Malmejac et al., 1947; Hortling and Pekkerinen, 1949), intraperitoneal (Porter, 1953a,b), intramuscular (Long, 1947; Samuels, 1951; Burstein, 1952), and subcutaneous (Long and Fry, 1945; Bisso, 1953; Speirs, 1953) injection.

The fact that adrenalin stimulates the pituitary-adrenal system, frequently demonstrated by indirect methods of investigation, is confirmed by the change in the concentration of corticosteroids in the urine and blood or in the rate of their secretion into the blood after administration of adrenalin. Prolonged or a single injection of adrenalin into dogs and cats quickens the secretion of corticosteroids into the blood flowing from the adrenal (Endröczi et al., 1958a). Intravenous infusion of adrenalin (0.12 mg/kg) for 1 h almost doubles the level of 17-hydroxycorticosteroids in the peripheral blood of dogs. Noradrenalin (0.017 mg/kg), under equivalent experimental conditions, does not stimulate adrenocortical function. When the above doses of catecholamines were given, there was some evidence of their influence on the hemodynamics (Hardwood and Mason, 1965).

The stimulant action of catecholamines on the function of the pituitary-adrenal complex is confirmed by experiments in which substances simulating their action were used. Administration of adrenomimetics such as amphetamine (Mikhailova, 1955; Nasmyth, 1955; Eskin et al., 1957; Vinogradskii, 1965), paredrine and ephedrine (Ohler and Wakerlin, 1951), and isopropylnoradrenalin (Rasin et al., 1970) causes activation of the adrenal cortex.

On the basis of experiments which demonstrated the stimulant action of catecholamines, Long (Long and Fry, 1945; Long, 1947, 1952) put forward a hypothesis that adrenalin of the adrenal medulla regulates the pituitary-adrenal complex. It was first postulated that adrenalin acts directly on the anterior pituitary (McDermott et al., 1950a,b; Long, 1952). Later, after evidence had been obtained of the important role of the hypothalamus in the mechanisms regulating the pituitary-adrenocortical system, the view was expressed that this amine exerts its effect at the hypothalamic level (Vogt, 1955; Long, 1956).

However, besides the many investigations whose results indicate that adrenalin and noradrenalin stimulate adrenocortical function, other work has yielded negative results. In early studies the absence of a stimulant effect was shown with the aid of indirect tests, e.g., on rats (Gordon, 1950) and rabbits (Colfer et al., 1950).

Later, many other workers failed to find any increase in the corticosteroid level in the peripheral blood after administration of adrenalin. For example, Raile et al. (1953) found an increase in the 17-hydroxycorticosteroid concentration in only 6 of 20 determinations in the blood of children with rheumatic diseases receiving a subcutaneous injection of adrenalin in a dose of 0.01 mg/kg body weight. Intravenous injection of adrenalin into healthy persons in doses of 1 to 3 mg did not lead to an increase in the blood 17-hydroxycorticosteroid concentration despite definite eosinopenia (Nelson et al., 1952). Similar conclusions were reached by other workers after tests on sick and healthy adults (Kinsell et al., 1956) and children (Kelly et al., 1952; Ely et al., 1954): the corticosteroid level in the peripheral blood was not raised although adrenalin induced eosinopenia and hyperglycemia. Even intravenous injection of this amine for 4 h, in either small (0.1-0.4 mg) or large (1-3 mg) doses, did not increase the 17-hydroxycorticosteroid concentration in the peripheral blood plasma (Sandberg et al., 1953).

Because in many cases it was impossible to find an increase in the corticosteroid level in the peripheral blood after administration of catecholamines, Long (1956) postulated that after injection of adrenalin not only the rate of secretion of corticosteroids, but also the rate of their consumption by the tissues, is increased. In one of his earlier papers Long (1952) points out that, besides its sympathomimetic effect, adrenalin has a well-marked action on metabolism (increasing the absolute level and rate of glycogenolysis in the liver and muscles), which in his opinion must lower the blood level of the adrenocortical hormones. According to the theory of Sayers and Sayers (1948) a low glucocorticoid concentration in the blood must in turn stimulate the function of the pituitary-adrenal complex. However, many workers were unable to observe any effect of adrenalin on metabolism of the adrenocortical hormones.

Sandberg et al. (1953), like earlier workers (Recant et al., 1950), found no increase in the excretion of 17-hydroxycorticosteroids and 17-ketosteroids, but when they gave cortisone and hydrocortisone in conjunction with adrenalin they found no change in the rate of disappearance of the injected hormones from the blood stream. They therefore concluded that adrenalin does not stimulate the secretion of corticosteroids and does not change their metabolism in the tissues. This conclusion was later confirmed by Bacchus (1954) in rats. His experiments, performed on liver slices, also do not support the view that this amine affects the metabolism of adrenocortical hormones.

The reason why some investigators found no increase in the blood corticosteroid level after parenteral injection of adrenalin is thus not that the rate of secretion of the adrenocortical hormones is equal to the rate of their utilization by the tissues but, in all probability, that adrenalin stimulates the pituitary-adrenal system under by no means all conditions. In fact, the negative results led many workers to consider that adrenalin is not the only, or an obligatory, stimulator of the function of the hypothalamo-hypophyseo-adrenal system (Ronzoni and Reichlin, 1950; Gershberg et al., 1950; Recant et al., 1950; Last et al., 1950; Meyer, 1953; Wolfson, 1953; Franksson et al., 1954; Guillemin, 1955; Hume, 1958; Goldfien and Ganong, 1962).

For instance, Gershberg et al. (1950) consider that adrenalin is essential only for rapid activation of the pituitary-adrenal complex in response to weak stimuli. If, however, the animal is exposed to severe or prolonged stimulation, the adrenal cortex is stimulated without adrenalin. Daily intravenous injections of adrenalin in doses of 5-10 μg/kg/min for 1 h into dogs led to the development of resistance to the drug after a few days. This "habituation" was manifested by disappearance of the eosinopenia which was observed during the first 1-2 days (Last et al., 1950). These results, in Last's opinion, are evidence against the view that adrenalin is the only physiological stimulator of prolonged ACTH secretion.

The logical continuation of the work just mentioned is the research of Guillemin (1955), who injected an adrenolytic into rats for 7 days and obtained adaptation of the pituitary-adrenal complex to adrenalin (SKF-501) and noradrenalin (SKF-688). Against this background adrenalin and noradrenalin, which in normal animals lower the ascorbic acid concentration in the adrenals, did not stimulate adrenocortical function. However, exposure under the same conditions to other types of stress (formalin or immobilization) led to a clear decrease in the ascorbic acid concentration in the adrenals of the rats. Guillemin (1955) concludes that the action of catecholamines is secondary, indirect, and that these substances are not humoral factors with a role in the regulation of pituitary-adrenocorticotrophic function.

The view that catecholamines play an essential role in the mechanisms controlling pituitary-adrenal activity likewise cannot be reconciled with the numerous facts pointing to the absence of correlation between the intensity of secretion of corticosteroids and of peripheral catecholamines. For example, during exposure to chronic forms of stress – infection, starvation, or malnutrition – an increase in ACTH secretion is found without stimulation of the adrenal medulla (Vogt, 1955).

Demedullation does not block the response of the adrenal cortex to stress (Gordon, 1950) and the adrenals of dogs continue to secrete 17-hydroxycorticosteroids into the peripheral blood after division of the splanchnic nerves in the same volume as the adrenals of intact animals (Pankov, 1961). Schapiro and Aragonas (1963) conclude that the autonomic nervous system has no direct effect on adrenocortical function, for no difference was found in the content of corticosterone and ascorbic acid in the innervated and denervated adrenals during exposure to weak stress.

Some curious results were obtained by Hungarian workers (Tigyi et al., 1959). By the use of formalin or cooling, i.e., the ordinary stimuli used to study the general adaptation syndrome and which have the effect of stimulating pituitary adrenocorticotrophic function, these workers found no change in the catecholamine concentration in the adrenals of rats. After the animals had swum for a long time in cold water (2°C) the catecholamine content in the gland actually showed a slight increase. Saito (cited by Tigyi et al., 1959) found no increase in the secretion of catecholamines by dogs cooled in this way so that their rectal temperature was lowered to 1.5-2°C.

Stimulation of the medial surface of the orbital cortex in cats is accompanied by a decrease in the adrenalin concentration in blood flowing from the adrenal (Euler and Folkow, 1958). Meanwhile, under similar experimental conditions, Endröczi et al. (1958b) observed an increase in pituitary-adrenal function. Even more convincing examples of dissociation of the secretion of catecholamine and corticosteroids were obtained by workers who determined adrenalin, noradrenalin, and adrenocortical hormones simultaneously. Franksson et al. (1954), for instance, showed that during operative stress the secretion of hormones of the adrenal medulla and cortex does not always follow a parallel course. Of 11 patients with a normal course of the operation and postoperative period, only 2 showed an increase in both 17-hydroxycorticosteroids and catecholamines in the peripheral blood. In the rest only glucocorticoid secretion was increased during the first 38 h.

After electrical stimulation of the ventral hypothalamus or orbital surface of the frontal lobe in dogs, the secretion of corticosteroids into blood flowing from the adrenal is increased, while the secretion of catecholamines is unchanged. Electrical stimulation of certain areas of the hypothalamus increases the secretion of catecholamines into the adrenal venous blood, whereas the rate of secretion of adrenocortical hormones is unchanged. Consequently, in these experiments the secretion of catecholamines was not parallel to the secretion of corticosteroids. These results, taken from the paper of Goldfien and Ganong (1962), led these workers to conclude that, in dogs at least, secretion of adrenal medullary hormones is not only not essential for ACTH secretion, but even if it does take place it does not act as a trigger factor for the stimulation of pituitary adrenocorticotrophic function.

The numerous facts irreconcilable with the view of the catecholamines as the principal regulators of the function of the hypothalamo-hypophyseo-adrenal system led even such ardent supporters of this hypothesis as Long and Vogt to revise their views to some extent as long ago as in the middle of the 1950s.

In a review of his investigations into the dependence of ACTH secretion on the influence of adrenalin from the adrenal medulla, Long (1956) states that loss of the ability of the adrenal medulla to secrete does not change the secretion of ACTH. However, he considers that under certain conditions the secretory power of the adenohypophysis is depressed by blocking the liberation of adrenalin. As Vogt (1955) writes, in a state of rest the adrenal medulla secretes hardly any hormones, whereas under the same conditions the adrenal cortex secretes fairly large quantities of corticosteroids. Adrenalin evidently participates in the regulation of pituitary adrenocorticotrophic function only under certain stress situations which are accompanied by an increase in the secretion of corticosteroids and also by a rapid and large outpouring of adrenalin as the result of stimulation of the splanchnic nerves (Vogt, 1955).

Catecholamines of the adrenal medulla are thus by no means always essential for the activity of the hypothalamo-hypophyseo-adrenal system in stress situations: some stress states require the participation of the adrenomedullary system only with no detectable response of the pituitary-adrenal complex, and vice versa (Euler, 1964).

Adrenolytics

The hypothesis that catecholamines play a decisive role in the regulation of pituitary-adrenal function likewise is not supported by the results of experiments in which adrenolytics were administered. Some workers claim that chlorpromazine, in doses of 2-50 mg/kg, inhibits or even completely blocks the response of the pituitary-adrenal system to stress (Ohler et al., 1956; Christy et al., 1956; Rezbek and Votava, 1956; Olling and De Wied, 1956; Sevy et al., 1957). Other investigators have concluded that chlorpromazine, in the doses mentioned above, stimulates ACTH secretion. These results were obtained in observations on man and experiments on monkeys, dogs, and rats (Georges and Cahn, 1953; Filk and Loeser, 1954; Holzbauer and Vogt, 1954; Nasmyth, 1955; Egdahl and Richards, 1956; Harwood and Mason, 1957; Vogt, 1964; Chatterjee, 1965). The activating effect of chlorpromazine was shown not only by indirect tests but also by determination of the corticosteroid content in the adrenals (Yuwiler et al., 1965), in the peripheral blood (Harwood and Mason, 1957), and in blood flowing from the adrenals (Egdahl and Richards, 1956).

Under the influence of chlorpromazine, the pituitary-adrenal response to stress, including to injection of adrenalin, was normal and was not inhibited (Holzbauer and Vogt, 1954; Nasmyth, 1955; Sapeika, 1959). Even administration of chlorpromazine in large doses (30 mg/kg) for several days did not change the secretion of corticosteroids in dogs in response to emotional or surgical stress (Ganong and Betz, 1960). Similar results were obtained by the use of Pacatal, which is related to chlorpromazine (Olling and De Wied, 1958).*

Jakobson and Hortling (1954) injected another adrenolytic –

*The effects of chlorpromazine on endocrine functions are surveyed by De Wied (1967).

dehydroergotamine – four times into rats in a dose of 12 mg/kg together with 0.2 mg/kg adrenalin and, using the eosinophil test, demonstrated the blocking effect of the former on the pituitary-adrenal system. The same result was obtained in experiments in which the lymphocyte count was used as an index of adrenocortical function (Siderius and Gaarenstroom, 1952). On the other hand, it has been shown that dehydroergotamine and dehydroergotoxin do not block the eosinopenia in man after administration of adrenalin (August and Gubner, 1949; Tepperman and Tepperman, 1950; Braunsteiner et al., 1951), while ergotamine usually stimulates the adrenals in rats and does not inhibit their function after injection of adrenalin (Gershberg et al., 1950; Ronzoni and Reichlin, 1950).

Excitation of the adrenal cortex is observed after injection of dibenzyline in a dose of 2 mg/kg into dogs (Hamilton, 1952). This sympatholytic inhibits the early eosinopenia and does not block the decrease in the eosinophil count 4 h after injection of adrenalin (Hamilton and Horwath, 1953). On the basis of their investigations in which they used the adrenomimetic paredrine and the adrenolytic dibenzyline, Ohler and Sevy (1953, 1955) ultimately conclude that endogenous adrenalin cannot play a primary role in the response of the pituitary-adrenal complex to stress.

In early investigations in which indirect tests were used to assess pituitary-adrenal activity, conflicting results were obtained regarding the effect of dibenamine on adrenocortical function (Tepperman and Bogardus, 1948; Ronzoni and Reichlin, 1950; Fortier, 1951b; Fritz and Levine, 1951; Sawyer, 1952; Sawyer and Pakerson, 1953). However, by determining the concentration of hormones in the blood, it was found that slow intravenous infusion of dibenamine (10 mg/kg 15 min before surgical stress) does not inhibit the secretion of corticosteroids in dogs. The developing increase in the rate of hormone secretion reaches a maximum, for intravenous injection of 25 units of ACTH cannot cause it to increase any further (Nagy et al., 1962).

These numerous experiments in which various adrenolytics were used thus did not reveal a decisive role of the catecholamines in the regulation of pituitary-adrenal function. The results are open to interpretation in various ways. The hypothesis can be accepted that catecholamines play a very minor role in the mechanisms of activation of the pituitary-adrenal system. However, it can also be accepted that the catecholamines do exert an influence on the pituitary-adrenal system by their action not only on α-, but also on β-adrenergic receptors [in the classification of Ahlquist (1948); see also the survey of Sly and Heimlich (1966)]. The sympatholytics used in the experiments described above, which block α-adrenergic structures, were therefore not always effective.

Some preliminary conclusions can be drawn from the facts described above. Most workers now firmly accept that adrenalin and noradrenalin, injected parenterally, stimulate adrenocortical function. The adrenalin of the adrenal medulla acts in the same way. However, it is obvious that the catecholamines of the adrenals are by no means always essential for stimulation of the hypothalamo-hypophyseo-adrenal system to take place.

How do the catecholamines influence the pituitary-adrenal complex and what is the relative importance of central and peripheral adrenergic structures in this process?

Ways in Which Catecholamines Influence the Pituitary-Adrenal Complex

Peripheral Catecholamines

Adrenalin evidently has no direct action on the adrenal cortex. In observations on the isolated adrenals of dogs perfused with blood from normal and hypophysectomized animals of the same species, Vogt (1951) found no increase in the rate of secretion of adrenocortical hormones after the addition of adrenalin to the perfusion fluid. The addition of ACTH, on the other hand, increased the rate of secretion. Rosenfeld (1955) found no increase in the rate of secretion of corticosteroids from the isolated calf adrenal after the addition of adrenalin and noradrenalin to the perfusion fluid. Negative results were also obtained on isolated dog adrenals (Cushman et al., 1966).

Adrenalin stimulates the adrenal cortex through the pituitary, so that this effect is not observed in hypophysectomized animals (Long and Fry, 1945; Vogt, 1947; Long, 1947; McDermott et al., 1950b; Gershberg et al., 1950; Pickford and Vogt, 1951). It has also been shown that 1 min after intravenous injection of adrenalin into rats the ACTH concentration in the blood is increased (Farrell and McCann, 1952). These findings were confirmed by other workers (Long, 1956). Similar results have been obtained in man: intravenous injection of adrenalin into healthy men and women in a dose of 0.5 μg/kg was followed by an increase in the blood ACTH concentration after 2-2.5 min (Vernikos-Danellis and Marks, 1962).

In their early papers many workers who studied the way in which adrenalin acts on the pituitary-adrenal system stated that during stress the adrenalin secreted from the adrenal medulla acts directly on the adenohypophysis. This view arose as the result of experiments in which the adenohypophysis was transplanted into the anterior chamber of the eye or beneath the capsule of the kidney (Long, 1947, 1952; McDermott et al., 1950a,b; Fortier, 1951a,b; Vogt, 1960). However, the doses of adrenalin used in these experiments were much higher than physiological. In addition, the injection of histamine into the graft was accompanied by a similar effect, indicating that the action of adrenalin under these conditions is nonspecific.

As the views regarding the leading role of the hypothalamus in the activity of the pituitary-adrenocortical system developed, the opinion formed that adrenalin acts through the hypothalamus. It was considered (Long, 1956; Vogt, 1951, 1955, 1957, 1960) that during stress the adrenalin liberated from the adrenal medulla reaches the hypothalamus by the humoral route, where it stimulates CRF secretion. However, there are several serious objections to this view of the direct role of the catecholamines formed in the adrenal medulla in the central regulation of activity of the hypothalamo-hypophyseo-adrenal complex.

In most investigations into the role of catecholamines in the regulation of endocrine functions the adrenalin or noradrenalin was injected parenterally (subcutaneously, intramuscularly, intraperitoneally, or intravenously) or the concentration of adrenalin in the blood was raised by stimulation of the splanchnic nerves. Now, in such experiments it is impossible to distinguish between the effect of the drug on the central chemically reactive structures and its action on the corresponding peripheral chemoreceptors. As an agent with high biological activity, however, adrenalin has a powerful action on peripheral adrenergic structures. Its direct action on tissues innervated by sympathetic nerves is well known and calls for no special proof (Burger and Dekker, 1967). The various regions of the central nervous system receive information on the wide range of different effects arising under the influence of this amine at the periphery (changes in the hemodynamics, metabolism, and so on) via the spinal cord. This flow of nervous impulses may also be the cause of CRF secretion and of subsequent stimulation of adenohypophyseal function.

It will be remembered that adrenalin (Weil-Malherbe et al., 1959) and noradrenalin (Weil-Malherbe, 1960), if injected parenterally, penetrate with difficulty through the blood-brain barrier. In these workers' experiments, 2 min after intravenous injection of radioactive adrenalin and noradrenalin the radioactivity in all parts of the brain was at the same level as in the plasma. The radioactivity in the hypothalamus became higher than that of the plasma only after 30 min. Yet it is well known that under extreme conditions the ACTH level rises within the first 1-2 min (Farrel and McCann, 1952; Long, 1956; Vernikos-Danellis and Marks, 1962; Vernikos-Danellis, 1964). If it is assumed that the catecholamines liberated from the adrenal medulla during stress participate directly in the central regulation of the pituitary-adrenal system, their level in the hypothalamus after intravenous injection ought to rise at least simultaneously with elevation of the ACTH level.

Consequently, the problem of the possible action of catecholamines formed at the periphery on the pituitary-adrenal system directly through the central adrenergic structures remained undecided despite many attempts to solve it one way or the other. Despite a search of the literature, the author has found no record of an investigation in which a strict differentiation was made by one means or another between the direct central action of the catecholamines and their effects on peripheral adrenergic structures.

The most adequate method for the solution of this problem, in my opinion, is to differentiate between the central and peripheral actions by a combination of division of the brain stem with pharmacological analysis. Transection at the mesencephalic level would prevent nearly all sensory impulses from reaching the region of the central nervous system higher than the level of section. Giuliani et al. (1961), who used this method, showed that adrenalin, if injected intraperitoneally into rats after mesencephalic section, no longer causes a decrease in the ascorbic acid content in the adrenals as it does in animals with an intact brain. However, quite wrongly, in my opinion, to analyze the role of the catecholamines these workers chose adrenalin, for it is quickly absorbed and destroyed by the peripheral tissues, and, at the same time, it passes with difficulty through the blood-brain barrier. Furthermore, no proof of penetration of the parenterally injected adrenalin above the level of transection of the brain was obtained in these investigations, and, more important still, there was no proof of its physiological action.

In my own investigations to study the role of central adrenergic structures in the regulation of the hypothalamo-hypophyseo-adrenal activity (Naumenko, 1963, 1965a, 1967c), to assess the functional state of the pituitary-adrenocortical system in guinea pigs I used the methods of determination of corticosteroids in the peripheral blood (Yudaev and Pankov, 1958). I recorded cortical potentials electroencephalographically in order to obtain evidence of the penetration and action of the substances used in the investigations, above the level of brain section. Finally, I did not use adrenalin or noradrenalin because of their inherent disadvantages but replaced them by adrenomimetics which are more suitable for these purposes. The use of substances simulating the action of biologically active substances formed in the body as participating in its vital activities is promising for another reason also: it enables the central or peripheral adrenergic structures to be stimulated to some extent selectively. I therefore deliberately chose three adrenomimetics which differ in their degree of central and peripheral sympathomimetic action.

Effects of Adrenomimetics on Guinea Pigs with an Intact Brain

Brief Pharmacological Characteristics of the Adrenomimetics Used. 1. Pipradol (Meratran), or α-(2-piperidyl)benzhydrol hydrochloride. This compound in doses of 1-10 mg/kg has only a slight effect on peripheral adrenergic structures (Brown and Werner, 1954), but it has a well-marked central effect (Brown and Werner, 1954; Liberman, 1958; Kucherenko, 1961; Mashkovskii and Ilyuchenok, 1961; Mekhedova, 1961). The mechanism of the central action of pipradol was studied in the following years. This compound was shown to affect adrenergic structures of the reticular formation of the mesencephalon and pons (Pastukhov, 1964; Il'yuchenok, 1965). Administration of pipradol was found to reduce the noradrenalin concentration in the brain stem (Sharov, 1968).

2. Amphetamine. This well-known adrenomimetic has a marked effect on both central and peripheral adrenergic structures. There is evidence that the mechanism of action of amphetamine is linked with its ability to stimulate the secretion of noradrenalin by the brain tissue (Fleckenstein and Burn, 1953; Fleckenstein and Stöckle, 1955; Muscholl, 1966; Carr and Moore, 1970; Lundborg and Waldeck, 1971), to potentiate the response to noradrenalin (Swamy et al., 1967), and to stimulate catecholamine synthesis in the brain tissue (Besson et al., 1969). Its use for the analysis of the role of adrenergic structures would accordingly appear to be highly promising.

3. Naphazoline (Sanorin) or 2-naphthyl-1-methylimidazoline, used as the nitrate. This is an adrenomimetic with mainly peripheral action. It is similar in its action to adrenalin but its vasoconstrictor effect is less marked but more prolonged (Trubitsyna, 1960).

Changes in Behavioral Responses and the EEG and ECG. From 15 to 20 min after subcutaneous injection of pipradol in a dose of 5 mg/kg or 5-10 min after injection of a dose of 10 mg/kg, distinct changes were observed in the behavior of guinea pigs. The animals became very active. Some of them turned their head and "marked time." Sometimes the head movements became "pendulum-like" in character. Other animals ran around the cage with their head low, sniffing at the floor. Another conspicuous feature was the marked increase in movements of mastication.

The behavior of the guinea pigs afer administration of 5-10 mg/kg amphetamine showed similar changes although the excitation developed rather more slowly and was not quite so strong as after administration of pipradol.

After administration of naphazoline (1-3 mg/kg), however, no change in the animals' behavior was observed. The experimental guinea pigs, like the controls, spent most of their time sitting in one place, huddled close to each other, and only occasionally did some of them move around the cage sniffing it.

Administration of pipradol and amphetamine produced well-marked EEG activation. After administration of pipradol this activation response was reflected in the sensomotor cortex as a low-amplitude fast rhythm; in the parietal region a low-amplitude

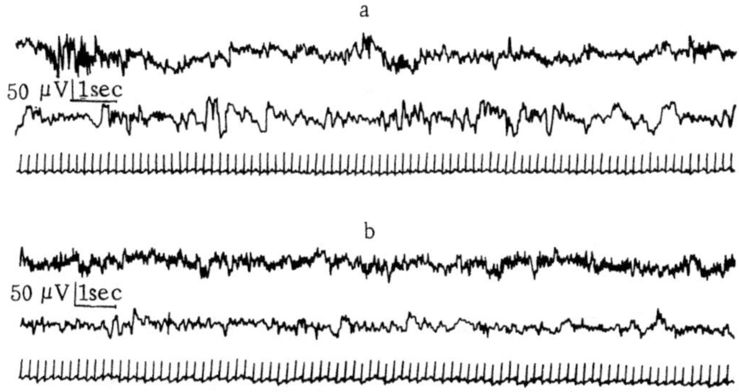

Fig. 1. Effect of pipradol on the EEG and ECG. From top to bottom: EEG of sensomotor and parietal regions of the cortex, ECG; a) before injection, b) 29 min after injection of pipradol (10 mg/kg).

but more regular rhythm appeared. No changes were found in the ECG (Fig. 1). After administration of amphetamine the changes in the EEG were similar. However, unlike pipradol, amphetamine produced changes in the ECG, consisting of bradycardia, changes in the voltage of the waves, and arrhythmia (Fig. 2). Naphazoline, in the dose (3 mg/kg) used, did not evoke an activation response. In some cases the EEG showed slow high-amplitude waves, probably due to stimulation of the baroreceptors and inhibition of the ascending activating system (Bonvallet et al., 1953). The changes in

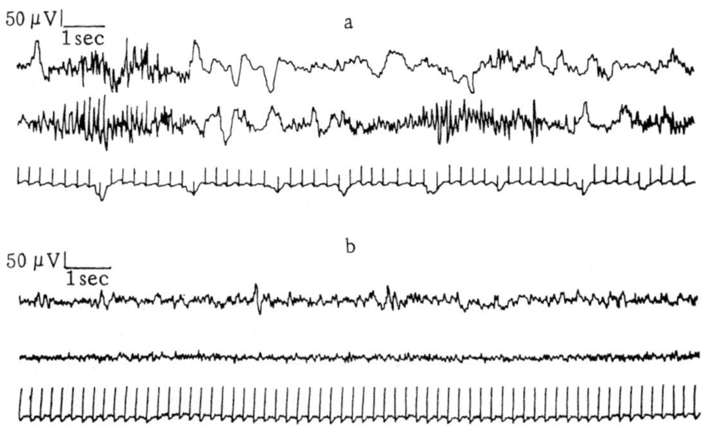

Fig. 2. Effect of amphetamine on EEG and ECG. From top to bottom: EEG of sensomotor and parietal regions of the cortex, ECG; a) before injection, b) 15 min after injection of amphetamine (5 mg/kg).

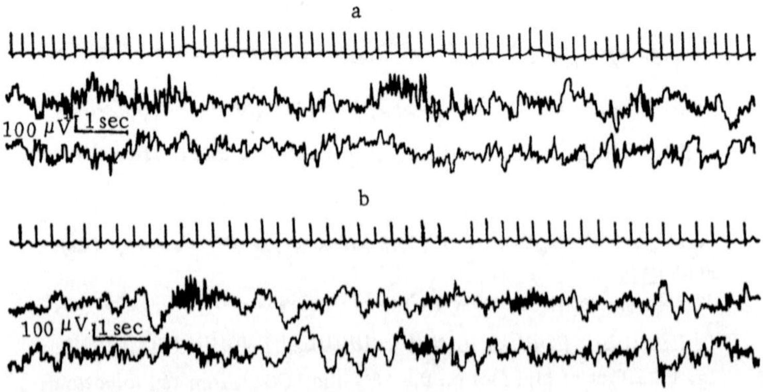

Fig. 3. Effect of naphazoline on EEG and ECG. From top to bottom: ECG, EEG of sensomotor and parietal regions of the cortex; a) before injection, b) 45 min after injection of naphazoline (3 mg/kg).

the ECG consisted mainly of bradycardia together with, in some animals, a slight reduction in voltage of the waves (Fig. 3).

Observations on the behavior and changes in the EEG and ECG of the animals thus confirmed results described previously in the literature: in the doses used, pipradol had a marked central stimulant effect, causing a sharp increase in the behavioral activity of the guinea pigs and activation of the EEG, while at the same time it had no visible effect on the peripheral adrenergic structures (the work of the heart was unchanged); amphetamine, with a distinct central effect, as shown by the changes in behavior and the EEG, at the same time had a marked peripheral action as reflected in the ECG; finally, naphazoline, with no visible effect on central adrenergic structures, led to changes only in the ECG.

If stimulation of the pituitary-adrenal system in fact takes place as a result of excitation of central adrenergic structures it would be expected that adrenomimetics would have an action on the adrenal cortex which corresponded in degree to their properties as a central stimulant. However, a study of pituitary-adrenal function yielded results diametrically opposite to those which would be expected if this assumption were correct.

Function of the Adrenal Cortex. The concentration of free 17-hydroxycorticosteroids in the peripheral blood plasma of guinea pigs was increased 1 h after subcutaneous injection of naphazoline.

The level rose progressively with an increase in the dose of naphazoline injected. Similar results were obtained after administration of amphetamine (Table 3).

The results reflecting the ability of amphetamine to activate the pituitary-adrenal system are in agreement with those obtained by other workers (Mikhailova, 1955; Nasmyth, 1955; Eskin et al., 1957; Vinogradskii, 1965). It is noteworthy that a marked stimulant effect was observed 1 h after injection of the drug, i.e., at a time when there was clear evidence of the adrenomimetic effects of amphetamine: EEG activation of the cortex associated with excitation of adrenergic structures of the brain-stem reticular formation (Il'yuchenok, 1965; Naumenko, 1965a) and maximal psychomotor stimulation (Dingel et al., 1967). It is interesting to note that 4 h after injection of the same dose of amphetamine (5 mg/kg) Bhattacharya and Marks (1969) found a decrease in the corticosterone level in the peripheral blood plasma and weakening of the response of the pituitary-adrenal system to stress. The decrease in reactivity of this system observed by these workers may be assumed to have taken place immediately after its activation by amphetamine. During activation of the pituitary-adrenal complex by amphetamine, inhibitory mechanisms are probably activated at the same time (elevation of the blood corticosteroid level is probably associated with this effect), and as their influence gradually increases in the course of time, it leads to a decrease in the blood corticosteroid concentration which may actually go below

TABLE 3. Blood 17-Hydroxycorticosteroid Level in Guinea Pigs with Intact Brain after Administration of Adrenomimetics

Substance injected	Mean plasma corticosteroid level ±m, μg/100 ml	P	Number of animals
Distilled water	36.8±4.32	<0.001	15
Naphazoline, 1 mg/kg	114.3±6.11		
Distilled water	47.1±6.88	<0.001	13
Naphazoline, 3 mg/kg	161.6±12.02		
Distilled water	41.5±4.68	<0.001	12
Amphetamine, 5 mg/kg	106.4±9.43		
Distilled water	42.5±3.52	<0.001	14
Amphetamine, 10 mg/kg	138.7±9.00		
Distilled water	37.3±4.90	>0.1	16
Pipradol, 5 mg/kg	44.9±6.42		
Distilled water	49.3±6.19	>0.1	31
Pipradol, 10 mg/kg	55.9±5.92		

the initial level. Against this background the stressor can no longer stimulate the pituitary-adrenal system effectively.

By contrast with the clearly defined activating effect of amphetamine and naphazoline, 1 h after subcutaneous injection of 5 mg/kg pipradol, the 17-hydroxycorticosteroid concentration in the peripheral blood plasma was increased in only 2 of the 16 animals After administration of the larger dose (10 mg/kg) of pipradol, no statistically significant increase in the corticosteroid level in the peripheral blood plasma of the guinea pigs likewise took place (Table 3). However, in 9 of 25 animals the blood hormone level was increased on the average by 121%. No relationship could be found between this increase and the intensity of muscular activity induced by the high level of excitation of the animals, although this possibility cannot be ruled out (Selye, 1952).

A special feature observed after the administration of pipradol and naphazoline was dissociation between the behavioral responses and function of the pituitary-adrenal system: marked excitation of the animals and a weak response of the adrenal cortex after administration of pipradol, and the opposite state of affairs after administration of naphazoline. It can be concluded from these findings that the administration of pharmacological agents producing excitation of an animal by no means always leads to stimulation of pituitary-adrenal activity. Dissociation between emotional responses and pituitary-adrenal function has also been observed by other workers (Mason, 1958; Lissak and Endröczi, 1964, 1965). The secretion of ACTH is considered to be determined not by the degree of motor activity but by its "emotional charge," by the affective component (Endröczi and Lissak, 1960).

In their account of a pharmacological study of pipradol, Brown and Werner (1954) emphasized that the dogs into which this substance was injected became more friendly toward each other despite their high level of excitation. Presumably pipradol induces excitation which is associated with a positive emotional response. No results of stimulation of the pituitary-adrenal system at a time of development of such "positive" emotions could be found in the literature. However, since this is a particularly interesting problem, it is worthwhile examining in more detail.

Experiments in which adrenomimetics with different actions on central and peripheral adrenergic structures were injected subcutaneously showed a direct linear relationship between the effect of the adrenomimetic on the corresponding peripheral adrenergic structures and the response of the pituitary-adrenal system; the stronger the peripheral action of the adrenomimetic, the stronger its stimulant effect on that system.

These results suggested that adrenomimetics act on the pituitary-adrenocortical system through excitation of peripheral structures. This hypothesis could evidently be confirmed by experiments in which the central action of a drug with a mixed effect on adrenergic structures could be differentiated from its peripheral effect. Of the group of substances studied, the only adrenomimetic with such properties was amphetamine. In addition, if the manifestation of peripheral sympathomimetic effects was rendered impossible, a drug exciting peripheral adrenergic structures should no longer induce activation of the pituitary-adrenal system. The drug with these properties used in the experiments was naphazoline. The next series of experiments was accordingly carried out on animals receiving amphetamine and naphazoline after mesencephalic brain section (Naumenko, 1965a, 1967c,d,e).

Action of Adrenomimetics on Guinea Pigs after Mesencephalic Section

Two types of mesencephalic section were used in the experiments: pretrigeminal, in which the brain stem was divided posterior to or through the inferior colliculi and immediately anterior to the point where the trigeminal nerves leave the pons, and the "cerveau isolé" (Bremer, 1935), in which the plane of section passed between the superior and inferior colliculi or through the superior colliculi and immediately anterior to the pons (Fig. 4).

If the stimulant effect of amphetamine and naphazoline on the hypothalamo-hypophyseo-adrenocortical system clearly observed in the previous series of experiments were a result of the direct

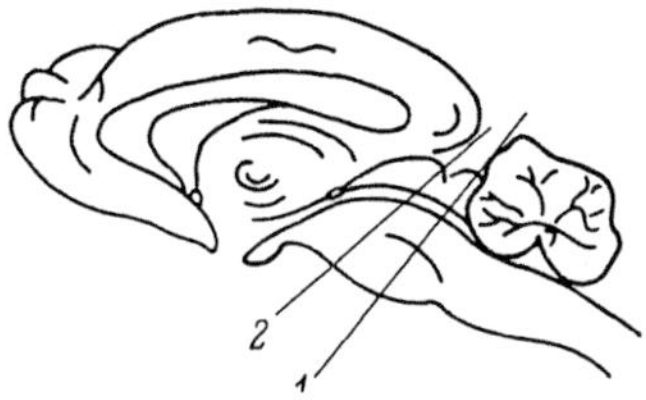

Fig. 4. Scheme of sections through the guinea pig brain: 1) plane of pretrigeminal section; 2) plane of section of the "cerveau isolé" type.

TABLE 4. Blood Level of 17-Hydroxycorticosteroids in Guinea Pigs Receiving Amphetamine and Naphazoline after Mesencephalic Section

Stage of investigation	Mean plasma corticosteroid level ±m, μg/100 ml	P	Number of animals
Initial level	66.3±4.40	>0.1	25
Amphetamine, 10 mg/kg . . .	71.9±4.98		
Initial level	59.8±3.88	>0.1	10
Naphazoline, 3 mg/kg	60.1±6.73		

central action of the adrenomimetics, these effects could be expected to remain in the mesencephalic section experiments. However, 1 h after subcutaneous injection of amphetamine in a dose raising the corticosteroid level in intact animals by 2.5 times, the blood corticosteroid concentration in guinea pigs with mesencephalic section was unchanged. This absence of response to amphetamine was found both after pretrigeminal section and in "cerveau isolé" animals, so that a single interpretation is possible for the results of these experiments. Similar results were obtained after administration of naphazoline (Table 4); these results clearly show that after mesencephalic section neither amphetamine nor naphazoline could stimulate the pituitary-adrenal system in the way which was observed in experiments on guinea pigs with an intact brain.

The absence of effect of amphetamine and naphazoline on the hypothalamo-hypophyseo-adrenal system in these experiments cannot be explained by the blocking of this system by brain section.

TABLE 5. Blood 17-Hydroxycorticosteroid Concentration in Guinea Pigs with Mesencephalic Section 1 h after Injection of Cold Physiological Saline into Lateral Ventricle

Stage of investigation	Mean plasma corticosteroid level ±m, μg/100ml	P	Number of animals
	Control		
1 h after section	71.7±6.21	> 0.1	12
2 h after section	74.3±6.15		
	Physiological saline		
1 h after section	62.3±4.31		
1 h after injection (2h after section)	122.4±7.20	< 0.001	25

In a special series of experiments, cold physiological saline (0.3 ml) was injected rapidly into the lateral ventricle 1 h after brain section. Definite stimulation of pituitary-adrenal function was observed 1 h after the injection (2 h after brain section) (Table 5). Further evidence was given by experiments in which serotonin was injected into the lateral ventricle under analogous conditions, when a similar effect was obtained (see Chapter IV).

These results likewise cannot be explained by a disturbance of the circulation in parts of the brain lying above the section or by the inability of amphetamine and naphazoline to penetrate above the level at which the brain was divided. After each experiment methylene blue was injected through the central end of the carotid artery and at autopsy distinct staining of the blood vessels of the entire brain was observed after all the experiments without exception.

Control experiments also showed conclusively, in the writer's opinion, that amphetamine under these conditions exerted its central action above the level of mesencephalic brain section. This conclusion is based on a study of cortical brain rhythms after brain section and before and after injection of the drug.

After injection of amphetamine, EEG activation was observed to develop in the sensomotor and parietal regions of the cortex, with the appearance of a low-amplitude fast rhythm. This change in the EEG was particularly marked after trigeminal section (Fig. 5).

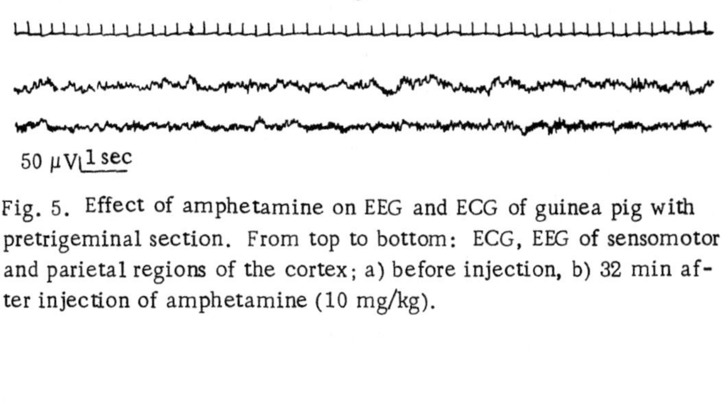

Fig. 5. Effect of amphetamine on EEG and ECG of guinea pig with pretrigeminal section. From top to bottom: ECG, EEG of sensomotor and parietal regions of the cortex; a) before injection, b) 32 min after injection of amphetamine (10 mg/kg).

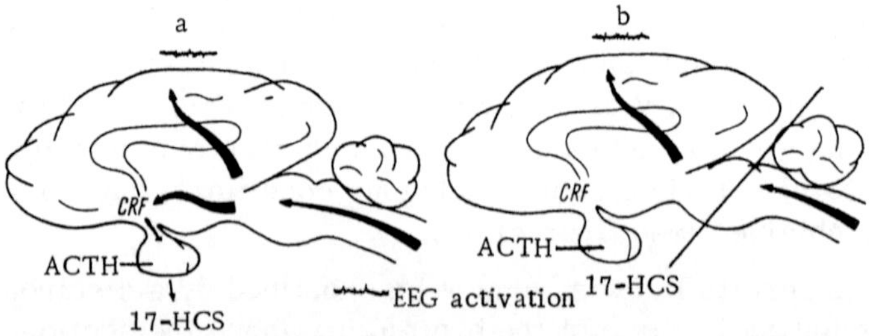

Fig. 6. Scheme showing effect of amphetamine on hypothalamo-hypophyseo-adrenal system. Subcutaneous injection of amphetamine is accompanied by excitation of central and peripheral adrenergic structures (a). However, the function of the pituitary-adrenal complex is stimulated only after excitation of peripheral adrenergic structures: after mesencephalic section, despite the central action of the drug in the brain above the level of section (EEG activation of the cortex), amphetamine has no activating effect on the hypothalamo-hypophyseo-adrenocortical system (b).

The activating effect of amphetamine persisted in the "cerveau isolé" preparation, but it was less marked than in the animals with trigeminal section, and it took the form of a decrease in amplitude and a slight increase in frequency of the rhythm. The fact that the "arousal" reaction was less marked but still clearly apparent in the "cerveau isolé" preparation can be explained by the isolation of the greater part of the mesencephalic reticular formation, on which amphetamine has a powerful action, in these experiments (Il'yuchenok, 1963).

The adrenergic structures of the mesencephalic reticular formation, a region lying above the level of section but closest to it, thus remain functionally active for they continue to respond by an EEG-activation reaction to injection of amphetamine under these conditions. Meanwhile, both amphetamine and naphazoline ceased to stimulate the function of the hypothalamo-hypophyseo-adrenal system. These experiments suggest that central adrenergic structures play no part in the stimulation of pituitary-adrenal function, or at least directly through the hypothalamus. The stimulant effect of amphetamine and naphazoline on this system in guinea pigs with an intact brain must evidently be regarded as the result of their primary effect on peripheral adrenergic structures (Fig. 6).

At the same time, it is impossible to disregard the few facts in the literature which suggest a role of the central adrenergic structures in regulation of the pituitary-adrenal complex. The next task was therefore to try to discover the reasons for disagreement between the conclusion drawn from the experiments described above and the admittedly indirect evidence suggesting a possible role of central adrenergic structures in the regulation of pituitary-adrenocortical function.

Role of Brain Catecholamines

The study of the role of the catecholamines formed in the central nervous system as mediators of nervous influences on the hypothalamus has been delayed principally by technical difficulties. This explains why, until recently, only indirect evidence, capable of different interpretations, has been available.

In a discussion of the role of a biologically active principle in the central regulation of pituitary-adrenal activity, it is natural to expect that the concentration of this substance in the central nervous system (or the rate of its metabolism) would change in situations of stress. As has already been pointed out, a high concentration of catecholamines (mainly noradrenalin) has been found in the hypothalamus, the gray matter around the aqueduct of Sylvius, the tegmentum mesencephali, the medial thalamus, and certain other regions (Vogt, 1957; Carlsson, 1959). Some workers have recorded a decrease, to a varied degree, in the brain noradrenalin content after stress following exposure to cold or electric shock in rats (Levi and Maynert, 1962; Maynert and Levi, 1964; Bliss et al., 1968) and monkeys (Ordy et al., 1966), after convulsions evoked by electric shock, Metrazol, or sound (Breitner et al., 1964), and also after operative trauma (Eskin and Shchedrina, 1964, 1966, 1968), head injury (Matlina et al., 1968), acute radiation sickness (Kulinskii, 1968), psychic stress (Corrodi et al., 1968; Welch and Welch, 1968), and stress induced in rats by horizontal oscillation (Rosecrans, 1969). Results of an opposite kind have also been found, when in response to various types of acute or chronic stress, an increase in the noradrenalin content was found in the rat brain (Bhagat, 1969; Riege and Morimoto, 1970). An increase in the rate of synthesis (Iwamoto and Sato, 1963; Gordon et al., 1966; Bhagat, 1969) or metabolism (Bhagat, 1969; Goldberg and Salama, 1969; Smookler and Buckley, 1969) also has been reported.

However, the results of such investigations can be interpreted in two ways: either these changes are in fact the cause of the change in function of the hypothalamo-hypophyseo-adrenal system or it is a parallel process, and the functional activity of this system changes as a result of the effect of other biologically active substances (such as acetylcholine or serotonin), the level of which also changes during stress (see below).

Endröczi et al. (1963) published the results of an investigation which showed that injection of adrenalin and noradrenalin through a microcannula into certain nuclei of the posterior hypothalamus and ventral part of the tegmentum mesencephali in cats evoked an increase in secretory activity of the hypothalamo-hypophyseo-adrenal system. Similar results were obtained by the Kriegers (D. Krieger and H. Krieger, 1965; H. Krieger and D. Krieger, 1970), who implanted noradrenalin crystals into the posterior hypothalamus and median eminence. Meanwhile, local administration of catecholamines into the anterior hypothalamus was not followed by changes in the rate of secretion of corticosteroids (Endröczi et al., 1963). These results suggested that the adrenergic structures of the posterior hypothalamus and rostral portion of the mesencephalon participate directly in the regulation of pituitary-adrenal function.

It is well known (Sager, 1960; Tonkikh, 1968), however, that the hypothalamus has some influence over the lower part of the brain stem and the spinal cord. Injection of substances raising the tone of the sympathetic nervous system into the posterior hypothalamus may give rise to sympathetic effects at the periphery, and this in turn may induce reflex stimulation of the pituitary-adrenal complex. Consequently, the experiments described above do not rule out the possibility of excitation of the corresponding mechanisms at the periphery through the transmission of impulses along efferent fibers in the spinal cord with a secondary response of the hypothalamo-hypophyseo-adrenal system. Since this is a problem of fundamental importance, the central effect of catecholamines injected into the brain had to be distinguished in these experiments from their indirect effect through the periphery. Experiments were accordingly carried out in which noradrenalin was injected directly into the brain through a microcannula inserted into guinea pigs with an intact brain and also after division of the brain stem (Naumenko, 1967a,c,d,e, 1968).

Effect of Local Injection of Noradrenalin into the Brain on Animals with an Intact Brain and with Mesencephalic Brain Section

The parts of the brain into which the cannula was inserted are shown schematically in Fig. 7.

After injection of noradrenalin into the various nuclei of the posterior hypothalamus and mesencephalon, no significant changes were observed in the guinea pigs' behavior. All that happened in some cases was a very slight increase in the animals' motor activity.

The response of the pituitary-adrenal system was unchanged 1 h after injection of physiological saline. Meanwhile, local injection of 1 μg noradrenalin into the nuclei of the mammillary complex and into the ventral tegmentum mesencephali led to a marked in-

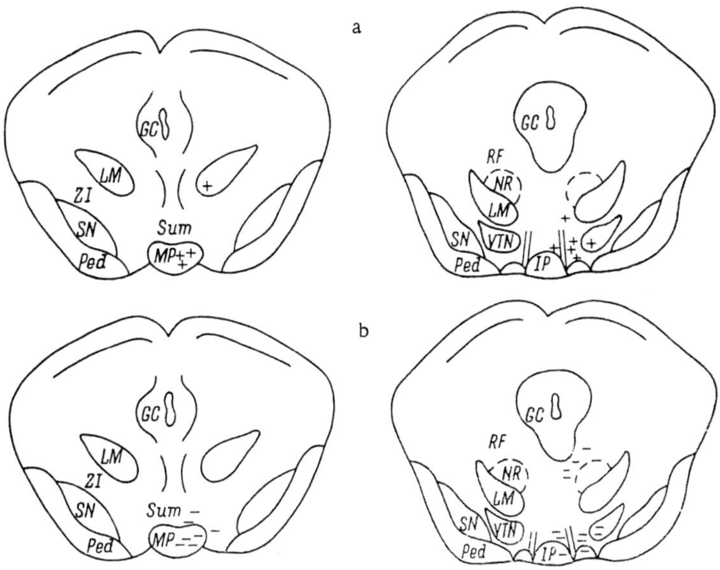

Fig. 7. Scheme of the posterior hypothalamus (left) and rostral portion of the mesencephalon (right): a) stimulant effect of noradrenalin (+) in guinea pigs with intact brain; b) no response to noradrenalin (-) after brain section. MP) Posterior mammillary nucleus; Sum) supramammillary region; SN) substantia nigra; Ped) cerebral peduncles; ZI) subthalamus; LM) medial lemniscus; GC) central gray matter; IP) interpeduncular nucleus; NR) red nucleus; RF) zone of mesencephalic reticular formation; VTN) ventral tegmental nucleus.

crease in the 17-hydroxycorticosteroid level in the peripheral blood plasma of the guineas pigs with an intact brain (Table 5). The results of these experiments on guinea pigs are in close agreement with those obtained in experiments on cats (Endröczi et al., 1963) and also with the results of the Kriegers' experiments (D. Krieger and

TABLE 6. Content of 17-Hydroxycorticosteroids in Blood of Guinea Pigs with Intact Brain and with Mesencephalic Section after Local Injection of 1 μg Noradrenalin

Localization of cannula	Mean plasma corticosteroid level ±m, μg/100 ml		P	Number of animals
	Original	Noradrenalin		
		Intact brain		
MP	72.4 ±1.47	127.8 ±9.50	< 0.01	3
MM	41.3	101.7		1
LM+Sum	49.2	82.7		1
VT	63.5 ±8.29	130.8 ±18.69	< 0.01	6
		After mesencephalic section		
MM	60.1	58.9		2
	64.2	61.7		
Sum	85.8 ±8.85	78.2 ±14.68	> 0.1	3
MP	63.6 ±5.82	55.9 ±6.22	> 0.1	4
PH	69.3	63.8		1
VT	75.4 ±6.26	67.4 ±6.42	> 0.1	8
RF	68.6 ±12.35	63.6 ±5.82	> 0.1	4

Note: MP) Posterior mammillary nucleus; Sum) supramammillary region; MM) medial mammillary nucleus; LM) medial lemniscus; PH) posterior hypothalamic nucleus; VT) ventral region of tegmentum mesencephali; RF) mesencephalic reticular formation.

H. Krieger, 1965). It is interesting to note that histochemical investigations (Dahlström and Fuxe, 1965a) have shown that catecholamine-containing nerve cells are located in those parts of the brain which stimulate the pituitary-adrenal system when noradrenalin is injected into them.

It has frequently been emphasized that experiments on animals with an intact brain can only yield evidence that excitation of certain central adrenergic structures is accompanied by stimulation of the pituitary-adrenal system. No information on the actual pathways transmitting this effect can be obtained from the results. The next series of experiments, along similar lines, was accordingly carried out under circumstances when all possible descending nervous influences were blocked.

When noradrenalin was injected into guinea pigs after mesencephalic section, no signs of activation of the hypothalamo-hypophyseo-adrenal system were observed, in contrast with animals with an intact brain. The plane of brain section in these experiments was naturally situated much more caudally than the regions into which the microcannulas were inserted.

Injection of noradrenalin into the medial and posterior mammillary nuclei, the submammillary region, the rostral portions of the ventral tegmentum mesencephali, or the mesencephalic reticular formation caused no change in the response of the hypothalamo-hypophyseo-adrenal system after 1 h. The blood corticosteroid level in the guinea pigs was unchanged 1 and 2 h after brain section (1 h after injection of noradrenalin). The results of this series of experiments are given in Table 6 and Fig. 7b.

Stimulation of the pituitary-adrenal system observed by other workers in cats and by myself in guinea pigs with an intact brain after intracerebral injection of noradrenalin is thus apparently explained by excitation of the adrenergic structures of the posterior hypothalamus and rostral portions of the mesencephalon, by subsequent transmission of impulses along the descending nerve fibers through the spinal cord, and by an indirect influence of the corresponding peripheral mechanisms on the pituitary-adrenal system (Fig. 8).

What are the possible mechanisms involved? It is well known that the hypothalamus is the integrative center for autonomic functions (Sager, 1960; Tonkikh, 1968). Excitation of the adrenergic structures of the hypothalamus is accompanied by a change in tone of the blood vessels, in the blood pressure, muscle tone, respiration, and body temperature (Share and Melville, 1963; Feldberg,

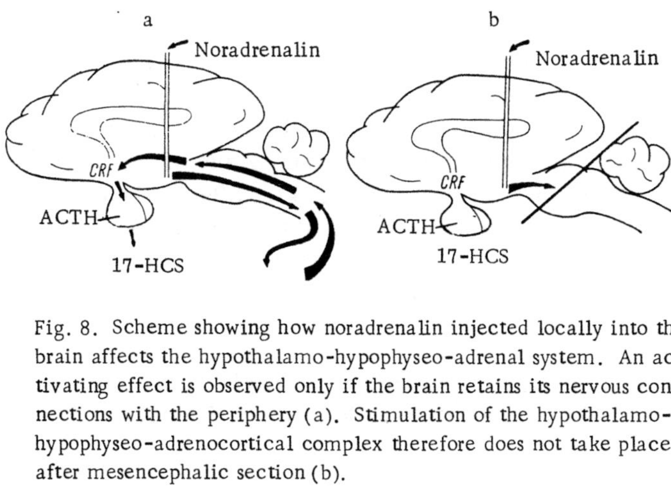

Fig. 8. Scheme showing how noradrenalin injected locally into the brain affects the hypothalamo-hypophyseo-adrenal system. An activating effect is observed only if the brain retains its nervous connections with the periphery (a). Stimulation of the hypothalamo-hypophyseo-adrenocortical complex therefore does not take place after mesencephalic section (b).

1966; Varagic et al., 1966). These responses are evidently sympathetic in nature and arise at the periphery through activation of the central adrenergic structures, and they may be the cause of the secondary stimulation of the hypothalamo-hypophyseo-adrenal complex (Ramey and Goldstein, 1957).

The study of the concrete mechanisms at the periphery is a special problem which is not within the scope of this book. However, evidence that such an indirect pathway for central adrenergic structures to act upon the hypothalamo-hypophyseo-adrenal system by the indirect route in principle is given not only by my results described above but also by the changes in homeostasis observed by other workers after excitation of hypothalamic adrenergic structures.

My experimental results suggest that only those adrenergic structures which belong to systems of descending neurons are connected with hypothalamo-hypophyseo-adrenal function. This may account for the absence of stimulation of the pituitary-adrenal system by noradrenalin injected locally into the brain after mesencephalic section and of amphetamine injected subcutaneously. The fact that pipradol does not act on this system does not contradict this hypothesis but may be explained on the assumption that this adrenomimetic does not affect the catecholamine structures of the brain connected with peripheral autonomic innervation. The possibility that adrenergic structures differing in their sensitivity to adrenomimetics exist in the brain cannot be ruled out. The existence of α- and β-adrenergic receptors, for example, is evidence for it (Ahlquist, 1948; Sly and Heimlich, 1966).

It is interesting to note that the results of some investigations have led the existence of afferent catecholamine pathways to the hypothalamus to be questioned. For instance, the noradrenalin level in the hypothalamus is not affected by any of the afferent fibers which terminate in it (Parent et al., 1969). Further evidence in this direction is given by experiments in which no change was found in the noradrenalin concentration in the hypothalamus after bilateral division of the medial lemniscus of the telencephalon (Benetato et al., 1967).

At the same time, it has been shown that the spinal cord has a high content of noradrenalin (Anderson and Holgerson, 1966).

Many noradrenalin-containing cells, with bodies in the brain stem, terminate there (Dahlström and Fuxe, 1965b). Some of these endings innervate primarily the sympathetic columns (Hillarp et al., 1966b). During stress, nerve endings containing noradrenalin are activated in the spinal cord (Corrodi et al., 1968).

It can accordingly be postulated on the basis of my experiments and observations of other workers that central adrenergic structures have no direct effect on the hypophyseal zone of the hypothalamus, producing CRF. Their activating effect on the hypothalamo-hypophyseal system is secondary and is transmitted via descending pathways through appropriate mechanisms at the periphery.

More recently, two further points of view have developed regarding the role of catecholamines in the regulation of the pituitary-adrenal system. Some workers have concluded that the brain catecholamines can participate in the regulation of this system, namely by inhibiting it (Lorenzen and Ganong, 1967; Van Loon et al., 1969, 1971a,b,c; Marks et al., 1970; Scapagnini et al., 1970). Other workers, disenchanted with the catecholamines as possible regulators of ACTH secretion, have concluded that central adrenergic neurons are not concerned with the regulation of the function of the pituitary-adrenocortical system during stress (Smelik, 1967; Carr and Moore, 1968; Hirsch and Moore, 1968). It is interesting to note that these two groups of workers carried out a very similar pharmacological analysis. They showed, for instance, that after intraperitoneal injection of α-methyl-p-tyrosine, an inhibitor of catecholamine synthesis reducing the content of noradrenalin and dopamine in the brain, the corticosteroid concentration in the peripheral blood plasma rises (Scapagnini et al., 1970; Van Loon et al., 1971c). A similar activating effect was observed after injection of α-methyl-p-tyrosine into the third ventricle (Van Loon et al., 1971c). In these workers' opinion their results indicate that the brain catecholamines may inhibit the pituitary-adrenocortical system. The same conclusion has been reached by other workers (Marks et al., 1970) who used reserpine in conjunction with monoamine oxidase inhibitors, inhibiting the response of the pituitary-adrenal system to stress, to investigate the role of adrenergic structures. Furthermore, preliminary injection of reserpine lowered the corticosterone concentration in the adrenals of rats receiving noradrenalin or dopamine by injection into the ventricles of the brain. The facts described above, in the opinion of the workers who obtained them, are evidence that the brain catecholamines may have an inhibitory effect on the pituitary-adrenal complex.

However, a decrease in the catecholamine concentration in the brain and a parallel increase in the blood corticosteroid level by themselves do not imply that the two processes are directly related to each other as cause and effect. A detailed analysis has shown that the lowered catecholamine level in the brain after injection of α-methyl-p-tyrosine and reserpine persists for a long time even after the blood corticosteroid level has returned to its initial value (Westermann et al., 1962; Carr and Moore, 1968). Moreover, emptying of the catecholamine depots by reserpine, whether injected locally into the region of the median eminence of the hypothalamus (Smelik, 1967) or after systemic administration (Carr and Moore, 1968; Hirsch and Moore, 1968), does not prevent the response of the pituitary-adrenal system to stress. The response to stress persisted, moreover, if reserpine was injected together with α-methyl-p-tyrosine, which more effectively blocked the participation of the catecholamines in neuronal activity (Carr and Moore, 1968). Furthermore, administration of reserpine together with monoamine oxidase inhibitors prevented the lowering of the noradrenalin and dopamine concentrations in the brain although it did not block the increase in the blood corticosterone level (Hirsch and Moore, 1968). From these results the authors in question concluded that the brain catecholamines do not play any essential role in the response of the pituitary-adrenal system to stress (Smelik, 1967; Carr and Moore, 1968; Hirsch and Moore, 1968).

In my opinion there is still insufficient evidence for such a categorical conclusion, for in the light of the facts described in this chapter it is impossible to rule out, in principle, the possibility that the brain catecholamines may participate in the activation of the pituitary-adrenal system by a system of indirect pathways through the periphery.

Nor can it be concluded from what has been said above that under special conditions excitation of central adrenergic neurons cannot inhibit the hypothalamo-hypophyseo-adrenal complex. In dogs anesthetized with pentobarbital during exposure to maximal stress (laparotomy), injection of large doses of noradrenalin, dopamine, amphetamine, isoproterenol, α-ethyltryptamine, and tyramine into the third ventricle is accompanied by a decrease in the 17-hydroxycorticosteroid concentration in the blood plasma in response to a second laparotomy (Lorenzen and Ganong, 1967; Van Loon et al., 1971b). These new and, in my opinion, very interesting facts obtained suggest that excitation of central neurons sensitive to catecholamines can, under special conditions, induce an inhibitory action on the pituitary-adrenal system. At the same time, we can agree with the conclusion drawn by the workers who performed these experiments that the physiological importance of their experiments and the role of central adrenergic neurons inhibiting ACTH secretion can be finally established only after further research.

Chapter III

Cholinergic Structures and the Regulation of Function of the Pituitary-Adrenal System

Acetylcholine as a Hypothetical Mediator in the Central Nervous System

Of the possible functions of chemical substances participating in the transmission of nervous impulses, it is the central mediator role of acetylcholine for which there is the most evidence. Although the actual nature of synaptic transmission in the brain is still uncertain, it is considered that the mechanisms by which acetylcholine participates in it are similar to the corresponding mechanisms of synaptic transmission in autonomic ganglia.

The weightiest evidence in support of cholinergic transmission in the central nervous system is given by experiments which showed that the Renshaw cells (interneurons in the ventromedial part of the anterior horns of the spinal cord, connected by synapses with fibers of motor axons and having a marked inhibitory action on motoneurons) are activated by acetylcholine liberated from ending of axon branches (Eccles et al., 1954; Curtis and Eccles, 1958; Curtis, 1961). Activity of the Renshaw cells is potentiated by choline esters, inhibited by cholinolytics, and prolonged by anticholinesterase substances. However, despite these facts, acetylcholine is not the only agent to excite Renshaw cells; other noncholinergic factors can also influence their activity, but their nature has not yet been established (Way and Sutherland, 1963). Presumably noradrenalin also plays some sort of role, mainly inhibitory (Weight and Salmoiraghi, 1966).

The mediator function of acetylcholine in the central nervous system is also confirmed by indirect evidence, the most important of which is as follows. The central nervous system contains certain specialized areas in which acetylcholine is synthesized, stored, and broken down. The highest content of acetylcholine is found in the brain stem (Feldberg, 1945; Malhotra and Pundlink, 1959). The uneven distribution of acetylcholine and of the enzymes concerned with its metabolism in the brain may indicate that its role varies in the function of different neurons. The highest content of choline acetylase, an enzyme catalyzing the last stage of acetylcholine synthesis, namely, the transfer of the acetyl group from acetylcoenzyme A to choline, is found in the motor cortex, thalamus, and geniculate bodies. Characteristically, cholinesterase (the enzyme catalyzing the hydrolysis of acetylcholine to choline and acetic acid) also is distributed in the brain in a manner parallel to the distribution of acetylcholine, but over a wider extent than choline acetylase (Way and Sutherland, 1963). The highest content of acetylcholinesterase is found in the caudate nucleus, the superior colliculi, and the lenticular nucleus (De Robertis and Fiszer, 1968).

Electron-microscopic investigations have located choline acetylase in presynaptic vesicles (De Robertis, 1961); acetylcholine has also been found in subcellular structures and also, evidently, in presynaptic vesicles (Whittaker, 1966). The presence of choline acetylase and acetylcholine in the presynaptic vesicles is particularly interesting, for there is reason to suppose (De Robertis, 1961) that these vesicles represent discrete units of mediator substance.

Numerous experiments using physiological and neurophysiological methods have demonstrated changes in the functional state of the central nervous system under the influence of acetylcholine and also changes in the synthesis, liberation, and breakdown of acetylcholine in relation to functional activity of the brain (Mikhel'son, 1948; Stone, 1957; Paton, 1958; Al'pern, 1963).

Much weighty evidence thus concentrates attention on acetylcholine as a possible mediator responsible for the transmission of impulses in the central nervous system. At the same time, the extent to which this role of acetylcholine is universal and, in particular, as it applies to our immediate problem, whether acetylcholine plays a direct role in the central regulation of the function of the pituitary-adrenal system, still remain open questions. In this chapter we shall examine the evidence which will enable us to answer this question.

Action of Acetylcholine and of Substances Exciting and Blocking Cholinergic Structures on the Pituitary-Adrenocortical System

There is no dispute about the fact that acetylcholine stimulates the pituitary-adrenal system (Guillemin, 1955; Casentini et al., 1957; Giuliani et al., 1961; Poskalenko, 1965). The results of experiments with acetylcholine are confirmed by those of other experiments using anticholinesterase drugs, i.e., substances leading to the accumulation of endogenous acetylcholine in the body. Dordoni and Fortier (1950) showed originally that intraperitoneal injection of eserine into rats leads to stimulation of adrenocortical function. This action is mediated through the pituitary, for hypophysectomy blocked the decrease in ascorbic acid concentration normally taking place after injection of this drug. Similar results were obtained by Poskalenko (1960). These results were also confirmed by experiments on unanesthetized dogs (Suzuki et al., 1964; Otsuka, 1966).

Besides anticholinesterase drugs, other substances stimulating adrenocortical function are drugs exciting cholinergic structures and, in particular, the muscarine-sensitive (M–) and nicotine-sensitive (N–) cholinomimetics. Activity of the pituitary-adrenal system is increased in response to injection of the M-cholinomimetics pilocarpine (Pol et al., 1957) and arecoline (Poskalenko, 1965), and this effect was observed in different species of animals. Administration of the N-cholinomimetic nicotine caused a decrease in the ascorbic acid concentration in the adrenals of rats after 10-20 min, but no such response was observed in hypophysectomized animals (Ryzhenkov, 1958).

Attempts have also been made to study cholinergic structures with the aid of substances blocking their function. Many pharmacological agents which block N-cholinergic receptors have nevertheless been found to stimulate the pituitary-adrenal system. Activity of the pituitary-adrenal system is potentiated by the Soviet N-cholinolytics diphacil (adiphenine), aprophen,* diprophen,† and tiphen‡ (Poskalenko, 1958, 1960, 1961, 1962, 1965; Tsoi, 1965).

*2-Diethylaminoethyl-2,2-diphenylpropionate hydrochloride.

†2-Dipropylaminoethyl diphenylthioacetate hydrochloride.

‡2-Diethylaminoethyl diphenylthioacetate hydrochloride.

The cholinolytics anicaine* and tropacin,† with mixed M- and N-cholinolytic action (Mel'nikova et al., 1958, 1959), atropine, which blocks peripheral M-cholinergic structures (Suzuki et al., 1964; Otsuka, 1966) also stimulate the adrenal cortex.

The use of ganglion-blocking drugs in the hope of determining the role of acetylcholine does not seem worthwhile. Although, for example, hexamethonium does not stimulate adrenocortical function (Compani and Peruzzo, 1953; Louwerens and Smelik, 1953; Arezio and Peruzzo, 1953; Poskalenko, 1965), the possibility cannot be ruled out that the absence of response is due to simultaneous blocking of both parasympathetic and sympathetic ganglia by this drug. It is accordingly difficult to interpret the results of such experiments.

Pathways for the Effect of Acetylcholine on the Pituitary-Adrenal Complex

Role of Peripheral Cholinergic Structures

The results of experiments performed by many different investigators have conclusively demonstrated that acetylcholine stimulates the pituitary-adrenal system. The question of how this effect takes place soon arose. The mode of action of acetylcholine is still not completely clear, although many interesting facts have been obtained in this direction, particularly in S. V. Anichkov's laboratory, and these will be mentioned later in this chapter.

Attempts have been made to discover whether acetylcholine can stimulate steroid production directly in the adrenals. Experiments (Rosenfeld, 1955) have shown that perfusion of the isolated calf adrenals with fluid to which acetylcholine is added increases the secretion of hydrocortisone, cortisone, and, to a lesser degree, corticosterone. Methacholine (β-methylacetylcholine) has a similar, although weaker, action. Carbachol (carbaminoyl choline), policarpine, and nicotine had no such effect. Rosenfeld concludes that the action of acetylcholine is connected with its specific structure and is independent of the possible effect of its hydrolysis products, choline and acetic acid. Nor is it connected with the ability of acetylcholine to stimulate the liberation of adrenalin and noradrenalin from the adrenal medulla: incubation of calf adrenals with

*2-Piperidinoethyl diphenylacetate.

†2,3-Dihydro-3-hydroxy-8-methylnortropidine diphenylacetate hydrochloride.

catecholamines did not affect steroid production. It is impossible from the results of the experiments described above to judge the hypothetical effect of acetylcholine on the adrenals in the intact organism, because Rosenfeld used higher concentrations of this substance (1-5 $\times$ 10^{-4}) than are found in the body.

Other workers (Macchi and Scotch, 1961) incubated rat adrenal glands with acetylcholine and found that this procedure led to the secretion of compounds which reduce tetrazolium blue. However, they did not observe any increase in the secretion of substances with a maximum of optical density in the region 240 nm. These workers consider that steroids of the Δ^4-3-keto group were secreted. They also found that the amount of secretion from the adrenals in the presence of acetylcholine was only one-fifth of the amount obtained on the addition of ACTH. Since acetylcholinesterase did not inhibit the action of ACTH, these workers conclude from their findings that ACTH acts on the adrenals through a mechanism which is not under the control of acetylcholine.

The direct effect of acetylcholine on the adrenals *in vivo* has been studied by transplantation of the glands into the neck of sheep (Blair-West et al., 1962). Injection of 50-120 μg acetylcholine per hour directly into the artery supplying the adrenal led in one experiment to an increase in aldosterone secretion, and in two experiments to an increase in the secretion of cortisol and corticosterone. However, parenteral injection of acetylcholine into rats after mesencephalic section did not lead to a decrease in the ascorbic acid content in the adrenals (Giuliani et al., 1961).

The investigations cited above thus failed to demonstrate precisely any direct effect of acetylcholine on steroid production and on the secretion of adrenocortical hormones, for the positive results of the experiments *in vitro* were not confirmed convincingly enough by the experiments on the intact animal.

Experiments carried out by S. V. Anichkov's collaborators are of great interest (Poskalenko, 1965; Ryzhenkov, 1958, 1959a,b). In experiments on dogs with a denervated and isolated carotid sinus, Ryzhenkov (1959a,b) showed for the first time that a decisive role in the action of nicotine on the pituitary-adrenal system is played by reflexes from the chemoreceptors of the carotid sinuses. The carotid sinus reflexogenic zone is known to be sensitive both to acetylcholine and to N-cholinomimetics (Anichkov and Belen'kii, 1962; Krylov, 1963). In chronic experiments on dogs injection of nicotine directly into the isolated carotid sinus stimulated adreno-cortical function, but if the nicotine was injected into the denervated sinus, no increase in the secretion of corticosteroids was observed. After bilateral denervation of the carotid bodies, intravenous injection of nicotine did not increase the 17-hydroxycorticosteroid concentration in the peripheral blood plasma (Ryzhenkov, 1959a,b). These results were confirmed by the work of Poskalenko (1965), who used acetylcholine in her investigations. She showed that for acetylcholine to exert its stimulant effect on the adrenal cortex, a local concentration of the drug high enough to produce excitation of the carotid sinus chemoreceptors must be created. I consider that her experiments in which acetylcholine was injected into the denervated carotid sinus of dogs through the common carotid artery, when the branches of the internal carotid artery were not ligated, are of fundamental importance. In this way the mediator could not reach the regions of the central nervous system supplied by its branches, notably the hypothalamic region (Harris, 1955). In such cases, however, stimulation of the adrenocorticotrophic function of the pituitary was not observed (in analogous experiments on rats Poskalenko found some decrease in the ascorbic acid concentration in the adrenals, but judging from the table, these changes were not significant). Consequently, denervation of the carotid sinus zone prevented the reflex from the cholinergic receptors to the hypothalamo-hypophyseo-adrenal system, thereby demonstrating the peripheral effect of acetylcholine on the pituitary-adrenal complex.

The subsequent pathways of the reflex action of acetylcholine are unkown. It may be that the hyperadrenalinemia arising after excitation of the carotid bodies (Petropavlovskaya, 1953) plays a role in this response. However, experiments in which acetylcholine was injected into the carotid body after division of the splanchnic nerves are against any participation of the catecholamines of the adrenal medulla in this reaction (Poskalenko, 1965).

Experiments in which acetylcholine is injected into the carotid bodies do not of course rule out the possibility that its effects on the pituitary-adrenal system may arise through excitation of other peripheral cholinergic structures, but they do provide convincing proof of the important role of these carotid structures in the mechanism of acetylcholine stimulation of the pituitary-adrenal system.

The Role of Central Cholinergic Structures

The role of central chemoreceptive structures in the regulation of pituitary-adrenal activity has not yet been explained. Two problems in particular confront the investigator: can acetylcholine, as a mediator of nervous impulses in the brain, influence the pituitary adrenal system and, if it can, how (directly or indirectly through the periphery) can the effects of excitation of the cholinergic brain structures influence the function of this system?

The role of acetylcholine as corticotrophin-releasing factor is not now accepted. Intraperitoneal injection of Mecholyl (acetyl-β-methylcholine) into rats is accompanied by a decrease in the ascorbic acid content in the adrenals. However, after administration of atropine for 6 days, at the end of which it did not stimulate adrenocortical function, injection of Mecholyl caused no change in the ascorbic acid content in the adrenals. Meanwhile, during prolonged administration of atropine, the use of psychic stressors and systemic stressors other than Mecholyl led to stimulation of the function of the pituitary-adrenal system (Guillemin, 1955). On the basis of these experiments Guillemin rejected acetylcholine as a corticotrophin-releasing factor. Casentini et al. (1957) concluded from their experiments in which they injected acetylcholine into normal and hypophysectomized rats and into animals with a pituitary gland grafted into the anterior chamber of the eye that acetylcholine cannot play the role of direct transmitter of nervous impulses from the hypothalamus to the adenohypophysis, although the intact hypothalamo-hypophyseo-adrenal complex is known to react to acetylcholine.

Although these results indicate that acetylcholine is not a corticotrophin-releasing factor, they do not disprove its role as a possible mediator of nervous impulses in the central nervous system stimulating secretion of CRF.

Indirect support for a possible role of the acetylcholine of the brain in the regulation of pituitary-adrenal function is given by the change in its concentration under different situations of stress. The early investigations of Rikkl' (1934) and Konradi and Mikhel'son (1935) showed that an acetylcholine-like substance appears in the blood flowing from the brain during stimulation of the central end of the vagus nerve. An increase in the acetylcholine concentration in fluid flowing from the cerebral cortex has also been observed during stimulation of other centripetal nerves (McIntosh and Oborin, 1953). The acetylcholine concentration in the brain rises during general anesthesia, whereas during excitation and in paroxysmal states its concentration falls (Richter and Crossland, 1949; Elliott et al., 1950; Crossland, 1953; Smirnov, 1957).

Other indirect evidence in support of the possible role of central cholinergic structures in the functional activity of the pituitary-adrenal system is given by experiments to study the action of the M-cholinomimetic arecoline which, according to Poskalenko (1965), stimulates this system by exciting central M-cholinergic structures.

The N-cholinolytics adiphenine, aprophen, diprophen, and tiphen have been shown to stimulate the pituitary-adrenal system by their action through the pituitary gland (Poskalenko, 1965). At the same time, adiphenine iodoethylate, its quaternary analogue which penetrates through the blood-brain barrier, does not change the adrenocortical function. On this basis Poskalenko (1965) concludes that the central N-cholinolytics act on the pituitary-adrenal system by blocking central N-cholinergic structures. Meanwhile, the experiments using adiphenine iodoethylate provide evidence against the participation of peripheral N-cholinergic structures in the stimulation of this system.

Finally, support for the role of central cholinergic structures in the regulation of the pituitary-adrenal system is given by the experiments of Endröczi et al. (1963): local injection of carbachol into the preoptic zone and the septal region in cats reduces the secretion of corticosteroids into the blood flowing from the adrenals. Meanwhile, injection of carbachol into the posterior hypothalamus and rostral portions of the mesencephalon is accompanied by stimulation of the function of the pituitary-adrenal system. D. Krieger (1965) implanted crystals of carbachol into the posterior hypothalamus and also observed its activating effect on the pituitary-adrenal system. Recently, in the course of a more detailed study of the localization of chemoreceptors connected with the function of the pituitary-adrenal complex in the brain, these same workers found that the corticosteroid level in the peripheral blood also rises after injection of carbachol into the median eminence and into certain parts of the amygdala (H. Krieger and D. Krieger, 1970). Similar results have been obtained in rats after injection of carbachol into the lateral ventricles (Marks et al., 1970). It can accordingly be postulated that the central cholinergic structures can participate in the regulation of activity of the pituitary-adrenal system. However, the pathways by which the corresponding chemically sensitive structures of the central nervous system exert their influence on the pituitary and adrenal cortex still remain unexplained, for, just as in the experiments in which noradrenalin was injected locally (Endröczi et al., 1963; 1963; D. Krieger and H. Krieger, 1965), the possibility that carbachol may act indirectly through the transmission of nervous impulses to the periphery via the spinal cord, followed by involvement of the pituitary-adrenocortical system in the response, cannot be ruled out.

Some of the workers cited above used acetylcholine to investigate the role of cholinergic structure. From my point of view, however, the use of acetylcholine in such investigations cannot be justified because of its rapid hydrolysis after injection. To analyze the role of cholinergic structures it is better to use carbachol, which differs from acetylcholine in its high resistance to cholinesterase, and substances leading to the accumulation of endogenous acetylcholine in the body. By using anticholinesterase drugs differing in their structure and also, consequently, in their penetration through the blood-brain barrier, an attempt can be made to differentiate between the response of the pituitary-adrenal system to excitation of central and peripheral cholinergic structures.

It is stated in the literature that the conversion of tertiary nitrogen into quaternary weakens the central and strengthens the peripheral cholinomimetic properties (Zeimal' et al., 1957; Zeimal' and Rybolovlev, 1957; Mashkovskii and Al'tshuler, 1962). This effect is considered to be attributable to the relative inability of quaternary ammonium compounds to penetrate into the central nervous system (Schweizer et al., 1939; Mikhel'son et al., 1961a,b).

I therefore used (Naumenko, 1965b,c, 1967c) two anticholinesterase drugs, one (galanthamine) a tertiary and the other (neostigmine) a quaternary inhibitor of cholinesterase.

Effect of Anticholinesterase Preparations on Guinea Pigs with an Intact Brain

A Brief Account of the Pharmacology of the Drugs Used. 1) Galanthamine (Nivalin). This substance was first isolated by Proskurina and Yakovleva (1952, 1955) from *Galanthus woronowi.* As a tertiary ammonium compound galanthamine can readily penetrate through the blood–brain barrier, and so inhibit not only the cholinesterase of the blood serum but also the acetylcholinesterase of the brain (Paskov, 1958, 1959; Irwin and Smith, 1960; Il'yuchenok, 1963; Nesterenko, 1966). Although it has a weaker effect on the EEG than eserine, in corresponding doses its action is more stable and much more prolonged (Mashkovskii and Il'yuchenok, 1961).

2) Neostigmine. This quaternary ammonium base penetrates poorly through the blood–brain barrier in the doses usually given and does not inhibit the brain acetylcholinesterase in cats or rabbits (Prozorovskii, 1960/1961; Il'yuchenok, 1963; Nesterenko, 1966). On the other hand, in doses as low as 0.05-0.1 mg/kg, neostigmine completely inhibits the blood plasma cholinesterase activity (Il'yuchenok, 1963).

In order to analyze the pathways responsible for the action of acetylcholine, besides using anticholinesterase drugs of different structure, it is also necessary to have a reliable control in order to monitor the central or peripheral action of the substance used. In my investigations I therefore recorded the EEG and ECG and determined the decrease in cholinesterase activity in the blood and brain (Panosyan, 1958).

Autonomic Responses and Electroencephalographic and Electrocardiographic Findings. From 20 to 30 min after subcutaneous injection of neostigmine (0.05-0.1 mg/kg) into guinea pigs, the animals responded with salivation, lacrimation, and twitching

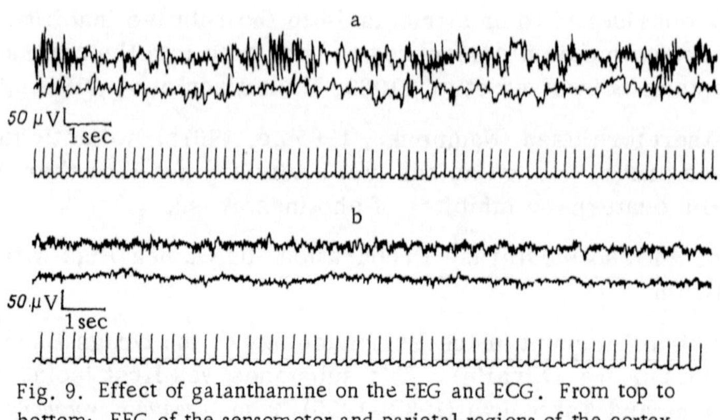

Fig. 9. Effect of galanthamine on the EEG and ECG. From top to bottom: EEG of the sensomotor and parietal regions of the cortex, ECG; a) before injection, b) 31 min after injection of galanthamine (5 mg/kg).

of individual muscle groups of the limbs, neck, and trunk. These responses developed progressively and became particularly marked 50-60 min after injection of the drug. Similar effects occurred after injection of galanthamine in a dose of 5-10 mg/kg.

Three minutes after injection of galanthamine the guinea pigs began to show a pattern of electroencephalographic activity similar to that observed after injection of this drug into cats and rabbits (Il'yuchenok, 1963, 1965). The high-amplitude slow waves in the sensomotor and parietal regions of the cortex were replaced by low-amplitude fast waves. Lengthening of the T–P interval, evidence of bradycardia, was observed on the ECG (Fig. 9).

Subcutaneous injection of neostigmine into the guinea pigs also produced marked EEG activation: the high-amplitude slow rhythm in the sensomotor and parietal cortex was replaced by a low-amplitude fast rhythm (Fig. 10). Although these changes, which were unexpected, developed gradually, only after 30-40 min, they were sufficiently well marked after the end of the first hour.

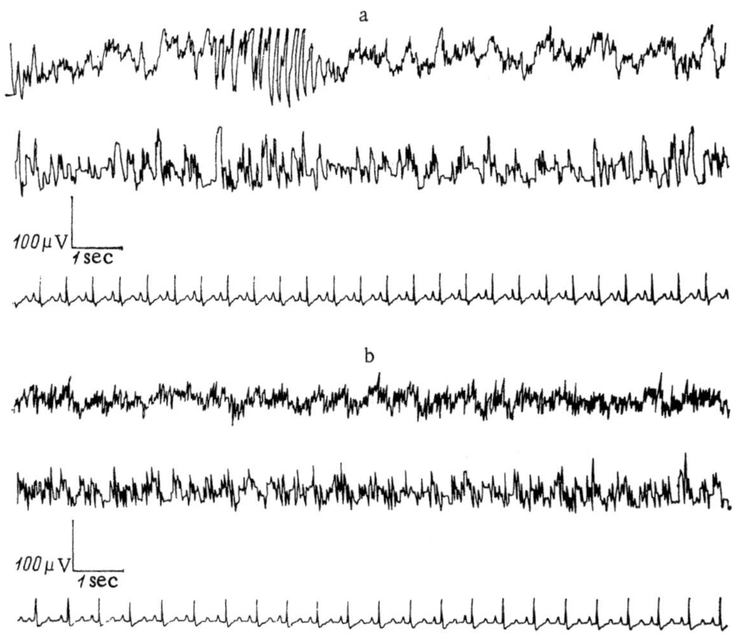

Fig. 10. Effect of neostigmine on the EEG and ECG. From top to bottom: EEG of the sensomotor and parietal regions of the cortex, ECG; a) before injection, b) 1 h after injection of neostigmine (0.1 mg/kg).

These results showed that neostigmine, although a quaternary ammonium derivative, can pass through the blood–brain barrier in guinea pigs, by contrast with what other workers have found in cats and rabbits (Il'yuchenok, 1963; Il'yuchenok et al., 1965; Nesterenko, 1966). This hypothesis was confirmed by experiments on guinea pigs in which the cholinesterase activity of the brain and blood serum was determined.

Changes in Cholinesterase Activity of the Brain and Blood. These experiments showed differences in the response of guinea pigs to administration of neostigmine and that of rabbits and cats. For instance, subcutaneous injection of neostigmine into guinea pigs in a dose as low as 0.0033 mg/kg was followed by a considerable decrease in acetylcholinesterase activity of the brain (Fig. 11). After this dose of the drug the activity fell on the average by 1.2 times below the control level; a more marked decrease in cholinesterase activity, by approximately 4.5 times, was observed in the blood.

An increase in the dose of neostigmine to 0.005 mg/kg reduced the acetylcholinesterase activity of the brain by approximately 1.5 times below the control level. Meanwhile, the blood cholinesterase activity fell to zero.

An increase in the dose of neostigmine to 0.05–0.1 mg/kg was not attended by any further statistically significant decrease in the brain acetylcholinesterase activity.

These results, confirming those obtained by recording the

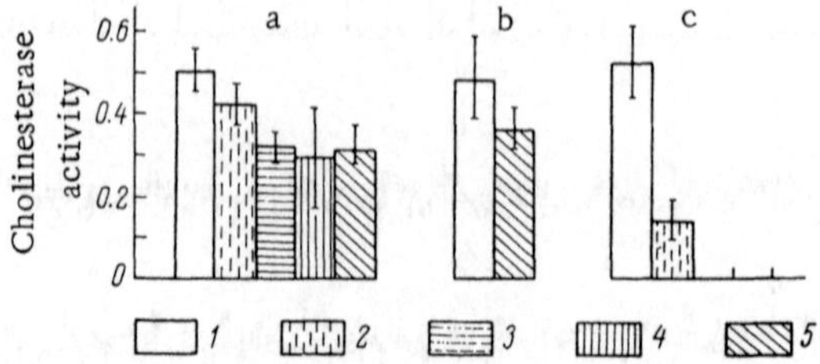

Fig. 11. Cholinesterase activity of brain and blood 1 h after injection of neostigmine (expressed in units of 0.001 mole acetic acid per 0.2 ml brain homogenate and per 0.1 ml blood serum): a) intact brain; b) section through brain stem; c) blood. 1) Normal; 2) 0.0033 mg/kg; 3) 0.005 mg/kg; 4) 0.05 mg/kg; 5) 0.1 mg/kg neostigmine.

EEG, showed that neostigmine passes through the blood–brain barrier relatively easily in guinea pigs. This is a very interesting fact, indicating as it does differences in the permeability of the blood–brain barrier to anticholinesterase drugs in animals of different species.

Subcutaneous injection of galanthamine in a dose of 2.5 mg/kg caused a decrease of 1.4 times in the brain acetylcholinesterase activity compared with the control (Fig. 12). Increasing the dose of the drug to 5 mg/kg body weight was accompanied by a further, although slight, decrease in brain acetylcholinesterase activity (by 1.7 times below the initial level). In a dose of 10 mg/kg, galanthamine lowered the brain acetylcholinesterase activity to approximately half its value in the control animals.

The inhibition of cholinesterase activity in the blood was of a different character. A dose of 2.5 mg/kg caused a sharp decrease in cholinesterase activity (to about one-quarter of the initial value). A further increase in the dose of the drug caused no significant change in the serum cholinesterase activity of the guinea pigs (the difference between the inhibition of blood cholinesterase activity after administration of galanthamine in doses of 5 and 10 mg/kg was not statistically significant).

Response of the Pituitary-Adrenal System. In a dose of 2.5 mg/kg, galanthamine did not cause a statistically significant change in adrenocortical function. An increase in the dose of the drug to 5 mg/kg approximately doubled the 17-hydroxycorticosteroid level in the blood of the guinea pigs. Finally, in a dose of 10 mg/kg, galanthamine led to a fourfold increase in the blood corticosteroid level compared with the control (Table 7).

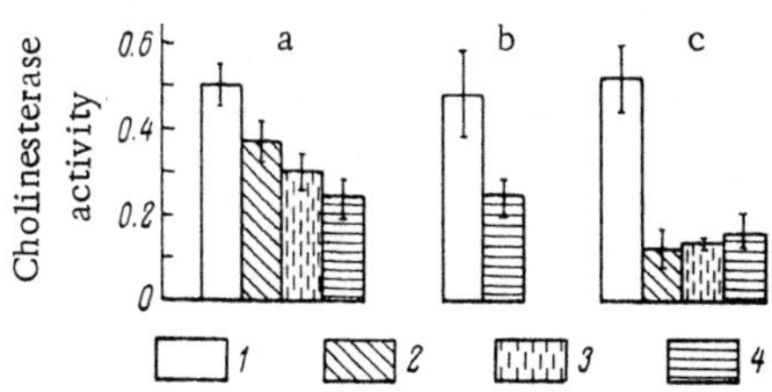

Fig. 12. Cholinesterase activity of brain and blood 1 h after administration of galanthamine: 1) normal; 2) 2.5 mg/kg; 3) 5 mg/kg; 4) 10 mg/kg galanthamine. Remainder of legend as in Fig. 11.

A study of the function of the pituitary-adrenal system after administration of neostigmine to the guinea pigs in doses of 0.0033 and 0.005 mg/kg showed that, in these doses, the drug did not increase the blood corticosteroid level. Administration of neostigmine in doses of 0.05 and 0.1 mg/kg was followed by a marked increase in the 17-hydroxycorticosteroid concentration in the peripheral blood, the level of which rose with an increase in the dose of the drug given.

It can accordingly be concluded from the results described above that in order to change the level of pituitary-adrenal function a certain critical degree of inhibition of cholinesterase activity is required or, in other words, a certain level of accumulation of acetylcholine in the body must occur.

Unfortunately, no conclusion regarding the role of peripheral or central cholinergic structures in stimulation of the pituitary-adrenal system could be drawn from the results of these experiments, for the blood corticosteroid level rose after administration of both neostigmine and galanthamine at a time when the cholinesterase activity was inhibited not only in the blood, but also in the brain. Consequently, the problem of differentiation between the central and peripheral action of these anticholinesterase drugs still remained unsolved. In view of these results it was necessary to carry out the next series of experiments, which, in my opinion,

TABLE 7. Blood 17-Hydroxycorticosteroid Levels in Guinea Pigs with Intact Brain after Administration of Anticholinesterase Drugs

Compound	Mean plasma corticosteroid level ±m, μg/100 ml	P	Number of animals
Distilled water	27.3±3.70	>0.1	17
Galanthamine, 2.5 mg/kg . . .	35.8±4.55		
Distilled water	35.2±3.94	<0.001	27
Galanthamine, 5 mg/kg	67.1±4.24		
Distilled water	26.5±2.84	<0.001	19
Galanthamine, 10 mg/kg . . .	97 2±6.92		
Distilled water	57.8±4.24	>0.1	10
Neostigmine, 0.0033 mg/kg .	51.8±4.24		
Distilled water	37.8±4.51	>0.1	10
Neostigmine, 0.005 mg/kg . .	51.1±10.38		
Distilled water	26.7±2.81	<0.001	12
Neostigmine, 0.05 mg/kg . . .	97.2±6.92		
Distilled water	33.4±3.93	<0.001	15
Neostigmine, 0.1 mg/kg	112.2±7.10		

provide a satisfactory means of differentiating between the peripheral effect of these drugs and their central action. I refer to the separation of these effects by mesencephalic brain section (Naumenko, 1965a, 1967d,e).

Action of Anticholinesterase Drugs after Mesencephalic Section

After subcutaneous injection of the maximal dose of galanthamine into guinea pigs with mesencephalic section, no statistically significant increase in the 17-hydroxycorticosteroid level in the peripheral blood plasma was observed (Table 8). This applied both to animals with pretrigeminal section and to "cerveau isolé" preparations.

The fact that there was no difference in the response of the pituitary-adrenal system after injection of 10 mg/kg galanthamine into guinea pigs with "cerveau isolé" and pretrigeminal brain section meant that the results could be regarded as synonymous. Equally negative results were obtained after administration of neostigmine in a dose of 0.1 mg/kg under analogous experimental conditions (Table 8).

The corticosteroid level in the peripheral blood of guinea pigs with mesencephalic section was thus unchanged 1 h after administration of galanthamine and neostigmine by comparison with its initial level determined 1 h after brain section and immediately before administration of the drugs.

This absence of response of the hypothalamo-hypophyseo-adrenal system after mesencephalic section cannot be attributed to blocking of its function, as was argued in detail in Chapter II. Moreover, the results of what I consider to be sufficiently objective control experiments indicate that these drugs, when injected sub-

TABLE 8. Blood 17-Hydroxycorticosteroid Concentration in Guinea Pigs Receiving Galanthamine and Neostigmine after Brain-Stem Section

Stage of investigation	Mean plasma corticosteroid level ±m, μg/100 ml	P	Number of animals
Initial level . . . . Galanthamine, 10 mg/kg . . .	75.1 ± 6.03 78.5 ± 6.93	>0.1	17
Initial level. . . . . Neostigmine, 0.1 mg/kg	78.4 ± 5.01 85.3 ± 7.43	>0.1	16

cutaneously, pass readily above the level of brain section and exert a clearly defined action:

1. Definite EEG activation was found in animals with both pretrigeminal section and "cerveau isolé" after administration of anticholinesterase drugs. High-amplitude slow waves (80-120 μV, 4-8 waves/sec) were replaced by a rhythm of 20-30 waves/sec, with an amplitude of 30-40 μV in the sensomotor and parietal regions of the cortex after injection of galanthamine in a dose of 10 mg/kg (Fig. 13). A similar picture was observed after subcutaneous injection of neostigmine in a dose of 0.1 mg/kg (Fig. 14).

The almost complete absence of an afferent inflow into regions of the brain above the level of section (with the exception of the first and second pairs of cranial nerves) did not change the pattern of the cortical EEG by comparison with that observed in guinea pigs with an intact brain. These observations show that the EEG activation observed after subcutaneous injection of galanthamine and neostigmine into guinea pigs is dependent on inhibition of brain acetylcholinesterase activity above the level of section and of excitation of cholinergic structures of the reticular formation which still remain functionally active.

2. After injection of galanthamine into guinea pigs, inhibition of acetyl-cholinesterase activity was observed above the level of section by about the same degree as in animals with an intact brain (Fig. 12). The cholinesterase activity in the blood serum was approximately 5 times less than in the control animals and indistinguishable from the serum cholinesterase activity of the guinea pigs with an intact brain. Similar results were obtained after administration of neostigmine: the subcutaneous injection of neostigmine in a dose of 0.1 mg/kg into these animals led not only to inhibition of acetylcholinesterase activity of the brain above the level of section, of the same degree as in the guinea pigs with an intact brain, but also to a fall in the serum cholinesterase activity to zero (Fig. 11).

3. Injection of methylene blue into the central end of the divided carotid artery in all experiments was followed by clear staining of the brain vessels above the level of section.

The activation of the cortical EEG, inhibition of brain acetylcholinesterase activity above the level of section, and the clear

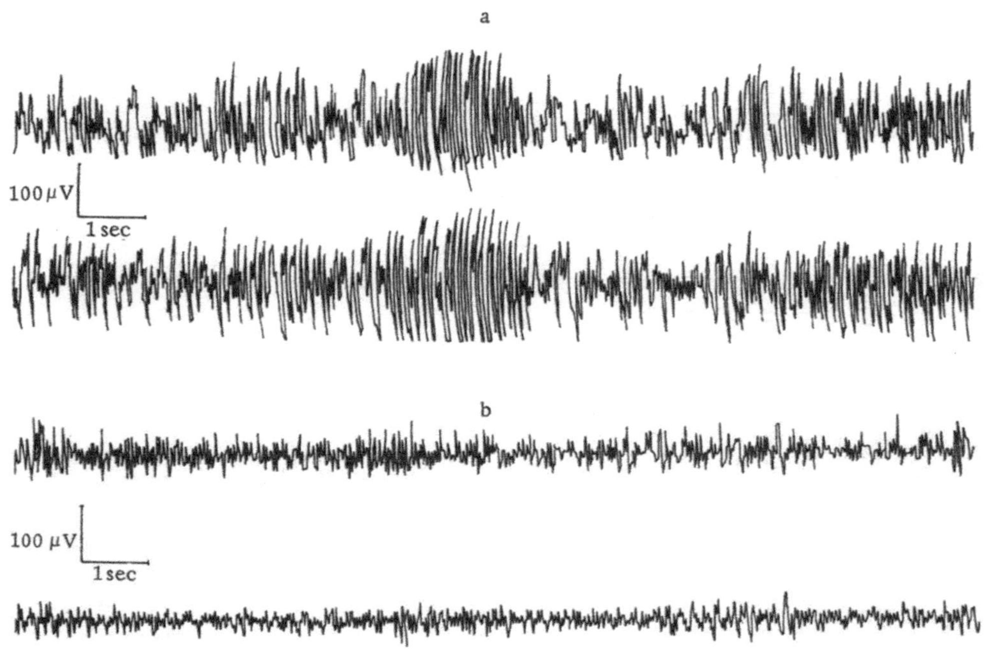

Fig. 13. Effect of galanthamine on the EEG after pretrigeminal section. From top to bottom: EEG of sensomotor and parietal regions of the cortex; a) before injection; b) 32 min after injection of galanthamine (10 mg/kg).

staining of all the brain vessels with methylene blue afford reliable evidence that the anticholinesterase drugs, when injected into guinea pigs after mesencephalic section, penetrated above the level of section and exerted a well-marked action. The absence of effect of galanthamine and neostigmine on the hypothalamo-hypophyseo-adrenal system after pretrigeminal section and in "cerveau isolé" preparations cannot therefore be explained by disturbance of the circulation in parts of the brain above the section.

The results of these experiments, together with those reflecting activity of the pituitary-adrenal system after mesencephalic section, can be interpreted as evidence against a direct stimulating effect of central cholinergic structures on the function of the hypothalamo-hypophyseo-adrenal system after their excitation (Fig. 15).

In fact, despite the increase in the brain acetylcholine concentration (inhibition of acetylcholinesterase activity), which was accompanied by excitation of central cholinergic structures (marked

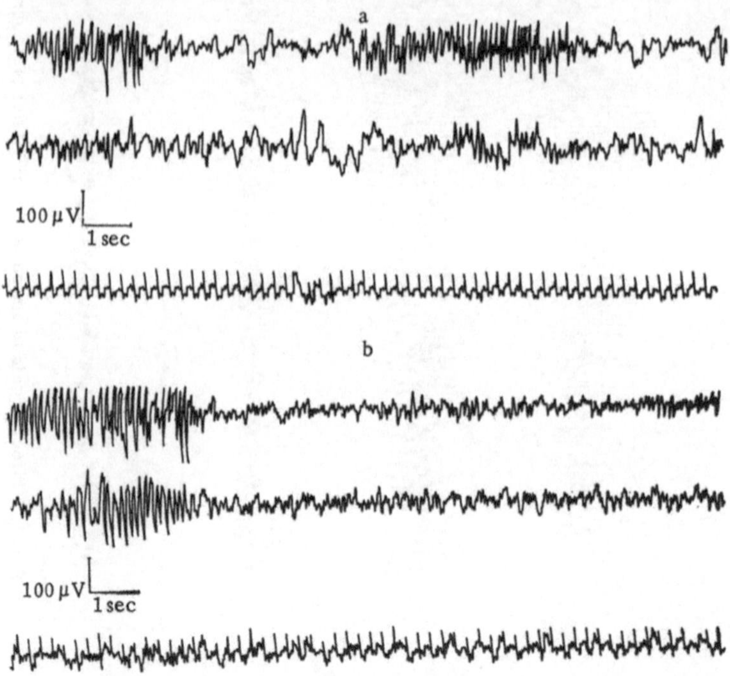

Fig. 14. Effect of neostigmine on EEG of guinea pig with "cerveau isolé." From top to bottom: EEG of sensomotor and parietal regions of the cortex, ECG; a) before injection, b) 1 h after injection of neostigmine (0.1 mg/kg).

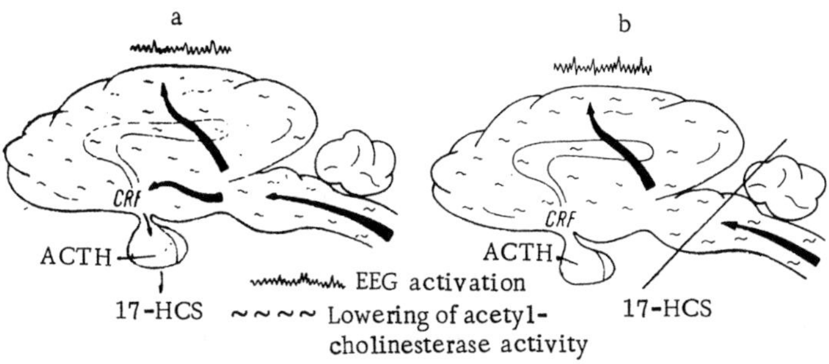

Fig. 15. Scheme of effect of anticholinesterase drugs on the hypothalamo-hypophyseo-adrenal system. Subcutaneous injection of galanthamine and neostigmine is followed by excitation of central and peripheral cholinergic structures (a). However, stimulation of the function of the hypothalamo-hypophyseo-adrenal complex is observed only after excitation of peripheral cholinergic structures: against the background of mesencephalic section (b), despite the central action of anticholinesterase drugs in the brain above the level of section (a decrease in acetylcholinesterase activity, shown on the diagram by short wavy lines, and cortical EEG activation), neither galanthamine nor neostigmine activates the hypothalamo-hypophyseo-adrenocortical system.

EEG activation), unlike in the animals with an intact brain, stimulation of the hypothalamo-hypophyseo-adrenal complex was not observed. This indicates the absence of a direct central effect of the cholinergic brain structures on the hypophyseal zone of the hypothalamus which secretes CRF. This conclusion was confirmed by experiments with local injection of carbachol into the brain of guinea pigs after mesencephalic section (Naumenko, 1967a, 1968).

Effect of Local Intracerebral Injection of Carbachol on Animals with an Intact Brain and after Brain Section

In the general scheme (Fig. 16) of regions of the posterior hypothalamus and the rostral portion of the mesencephalon, those areas into which carbachol was injected in experiments on guinea pigs with an intact brain (a) and after brain section (b) are shown. Carbachol (5 μg) was injected into the posterior hypothalamic nucleus, the posterior mammillary nucleus, into the region of the posterior hypothalamus lying above them, and into the ventral tegmentum mesencephali.

A few minutes after injection of the drug into the guinea pigs with the intact brain, their behavior changed sharply. They became very excitable and ran about the cage from place to place. As the experimenter approached, the restless movements were intensified. Among the autonomic responses, lacrimation and salivation were particularly marked.

The 17-hydroxycorticosteroid level in the peripheral blood plasma was distinctly raised 1 h after injection of 5 μg carbachol into the areas of the posterior hypothalamus mentioned above and into the rostral part of the mesencephalon (Table 9).

The results of these experiments show that in guinea pigs, just as in cats, local injection of carbachol into the posterior hypothalamus and rostral portion of the mesencephalon is accompanied by stimulation of adrenocortical function. The well-marked autonomic responses which were observed suggest that excitation of central cholinergic structures induces a subsequent response

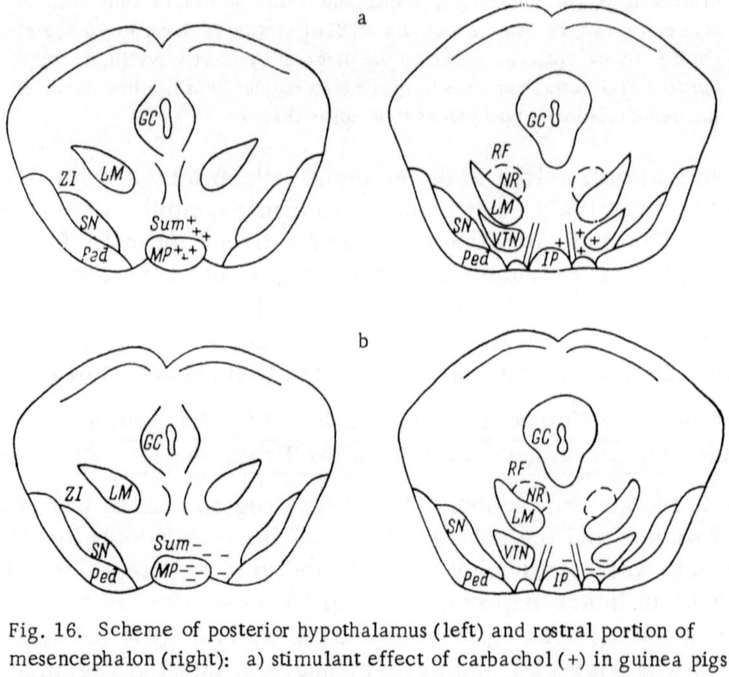

Fig. 16. Scheme of posterior hypothalamus (left) and rostral portion of mesencephalon (right): a) stimulant effect of carbachol (+) in guinea pigs with an intact brain; b) no response to carbachol (-) after mesencephalic section. MP) Posterior mammillary nucleus; Sum) supramammillary region; SN) substantia nigra; Ped) cerebral peduncles; ZI) subthalamus; LM) medial lemniscus; GC) central gray matter; IP) interpeduncular nucleus; NR) red nucleus; RF) zone of mesencephalic reticular formation; VTN) ventral nucleus of tegmentum.

TABLE 9. Blood 17-Hydroxycorticosteroid Level of Guinea Pigs with an Intact Brain and with Mesencephalic Section after Local Injection of 5 μg Carbachol

Localization of cannula	Mean plasma corticosteroid level ±m, μg/100 ml		P	Number of animals
	original	carbachol		
		Intact brain		
Sum	48.8 ±4.50	90.8 ±6.06	< 0.01	3
MP	63.2 ±5.16	109.5 ±6.55	< 0.01	3
PH	54.1	89.6		1
VT	62.3 ±3.35	103.5 ±5.87	< 0.002	4
		Brain section		
ML	74.6	76.6		2
	84.0	84.1		
Sum	85.0 ±7.17	75.1 ±8.62	> 0.1	4
MP	75.1 ±8.62	71.7 ±9.33	> 0.1	4
VT	82.4 ±2.34	80.3 ±5.92	> 0.1	4

Note. MP) Posterior mammillary nucleus; Sum) supramammillary region; PH) posterior hypothalamic nucleus; ML) lateral mammillary nucleus; VT) ventral tegmentum mesencephali.

at the periphery. However, to differentiate between the central and peripheral effects of carbachol when injected locally, similar experiments would evidently have to be carried out under conditions preventing the possibility of peripheral responses. This could be done by the use of a model with mesencephalic section. The plane of section must naturally pass caudally to the cannulas inserted for injecting the drug.

It has already been stated that the brain section in these experiments itself led to only a slight increase in the corticosteroid concentration in the peripheral blood plasma, which remained far below its maximal level. If, after brain section, carbachol was injected into the posterior mammillary nucleus and into the region above it, or into the ventral tegmentum mesencephali, the blood corticosteroid level of the guinea pigs did not rise (Table 9).

The results of these experiments were similar to those obtained after the local injection of noradrenalin. They indicate that the stimulant action of carbachol on animals with an intact brain is connected with excitation of the cholinergic structures of the posterior hypothalamus and rostral portion of the mesencephalon, with a subsequent response through the spinal cord at the periphery.

On the other hand, these experiments demonstrate the absence of ascending cholinergic pathways to higher levels of the brain concerned with the regulation of the pituitary-adrenal complex.

Data described in the literature and the results of my own investigations thus show that stimulation of the pituitary-adrenal system is observed after excitation of peripheral cholinergic structures. Endogenous acetylcholine, accumulating in the body after administration of anticholinesterase drugs, has no direct stimulating effect on the hormonal function of the adrenal cortex. Finally, the facts established give grounds for rejecting the direct participation of cholinergic brain structures in the activation of pituitary adrenocorticotrophic function. Their stimulant effect on the hypothalamo-hypophyseo-adrenal complex is indirect and is mediated through corresponding peripheral mechanisms.

Chapter IV

Serotoninergic Structures and Regulation of Pituitary-Adrenal Function

Serotonin (5-hydroxytryptamine), which was discovered comparatively recently, has attracted considerable attention (Page, 1954; Cerletti, 1958; Lewis, 1958; Quivy, 1959; Mogilevskii, 1960; Chernov, 1960; Costa, 1960; Garattini and Valzelli, 1965; Planel'es and Popenenkova, 1950; Gromova, 1966). However, its biological role is not yet established. This applies in particular to the role of serotonin in the activity of the central nervous system, a problem which has received the closest study.

Research in this direction has been stimulated in particular by the discovery of two principal facts: the presence of serotonin and of enzymes catalyzing its synthesis and breakdown in the brain, and the discovery that the two principal groups of psychotropic drugs, with opposite central effects, viz., monoamine oxidase inhibitors and reserpine, also have opposite actions on the tissue serotonin depots.

The scope of this book does not permit a detailed examination of the many different aspects (biochemical, physiological, and pharmacological) of the general problem of the action of serotonin. I shall therefore simply mention some of its aspects which illustrate the possible mediator role of serotonin in the central nervous system.

Serotonin as a Hypothetical Mediator in the Central Nervous System

Extensive observations have shown that serotonin is present in relatively large concentrations in the brain, especially in the pineal body, hypothalamus, amygdala, and caudate nucleus. Its concentration in the cerebral cortex and cerebellum, on the other hand, is low (Paasonen et al., 1957; Crawford, 1958; Davison, 1958; Garattini and Valzelli, 1965).

Exogenous serotonin penetrates with difficulty through the blood–brain barrier (Udenfriend et al., 1957a). The suggestion has therefore been made that the serotonin present in the brain under natural conditions was evidently formed there. Whatever the case, the brain has a high content of the enzyme tryptophan decarboxylase, which catalyzes the last stage of serotonin synthesis from 5-hydroxytryptophan, and administration of exogenous 5-hydroxytryptophan produces a rapid and considerable rise in the serotonin concentration in the brain (Udenfriend et al., 1957b; Bogdanski et al., 1958a,b; Bertler and Rosengren, 1959; Erspamer and Bertaccini, 1962). Tryptophan, from which 5-hydroxytryptophan is formed, is also found in the structures of the central nervous system, and its content in the hypothalamus is very high. Evidence has also been obtained that the brain contains tryptophan oxidase, an enzyme catalyzing the hydroxylation of tryptophan with the formation of 5-hydroxytryptophan (Consolo et al., 1965; Green and Sawyer, 1966; Lovenberg et al., 1967; Airaksinen et al., 1968).

The localization of monoamine oxidase, an enzyme of oxidative deamination – the principal pathway of serotonin breakdown – is parallel to the distribution of serotonin in the central nervous system (Bogdanski and Udenfriend, 1956; Davison, 1958; Weiner (1960).

Substantial support for the hypothesis of the possible role of serotonin as a neurotransmitter in the central nervous system was obtained after administration of the compound by microelectrophoresis. The results showed that many regions of the brain contain neurons sensitive to serotonin,* and some evidence was obtained that serotonin can act on synapses (Salmoiraghi et al., 1965).

Serotonin-sensitive nerve cells are diffusely distributed throughout the anatomical regions of the hypothalamus. They have

*An investigation of considerable interest in this connection is that of Woolley and Gommi (1966), who, by chemical fractionation, isolated from serotonin-sensitive cells "serotonin receptors" which proved to be special lipids, and in this way they established a biochemical basis for the hitherto hypothetical serotoninergic structures.

also been found in the thalamus and the lateral diencephalon (Bloom et al., 1963).

The use of a fluorescence method showed that serotonin is present in very high concentrations in synaptic endings in the central nervous system of warm-blooded animals (Dahlström and Fuxe, 1965a,b; Hillarp et al., 1966a,b), and the characteristics of "serotonin" neurons were described. These cells are usually small, oval or circular in shape. The cytoplasm around the nucleus gives very weak fluorescence, evidence of the presence of serotonin, but this becomes much stronger through the action of monoamine oxidase inhibitors. The nuclei of these cells do not fluoresce.

"Serotonin" neurons are found in the caudal part of the brain stem and in the central and medial parts of the reticular formation. The highest concentration of serotonin neurons has been found in the region of the raphe nuclei (Dahlström and Fuxe, 1965a,b) – a group of nuclei stretching throughout the mesencephalon and medulla. Serotonin-containing cells in the caudal ventromedial part of the medulla give off axons into the spinal cord (Dahlström and Fuxe, 1965b). The numerous serotonin cells of the midbrain raphe nuclei give rise to an ascending neuronal system which passes through the lateral hypothalamus in the medial forebrain bundle and is projected to hypothalamic and limbic structures (Anden et al., 1965; Heller and Moore, 1965; Gumulka et al., 1969; Sheard and Aghajanian, 1969; Holman and Vogt, 1970). Electrical stimulation of the medial midbrain raphe nuclei leads to an increase in the rate of serotonin metabolism in the forebrain (Gumulka et al., 1969). Axons of the serotonin neurons are similar in their general characteristics to those of the "noradrenalin" cells. Close contacts between endings of serotonin neurons and nonfluorescent (and, consequently, "nonserotonin") cell bodies have been found. The authors concerned (Hillarp et al., 1966a,b) consider that all these findings, together with those described above in the section on catecholamines, which apply in a large measure to serotonin also, afford weighty evidence of the existence of monoamine and, in particular, serotonin mediator activity in the central nervous system.

A link between adrenal function and serotonin (5-hydroxytryptamine) metabolism has been postulated, although the concrete mechanisms of this link have not yet been established. The literature on this question is not large and in many of its details it is contradictory.

The Adrenal Cortex, Stress, and Serotonin

Adrenalectomy

Evidence of a link between serotonin metabolism and adrenocortical function is provided by experiments showing changes in serotonin metabolism after removal of the adrenals. Such experiments (Garattini et al., 1961) showed that adrenalectomy in rats and mice increases the toxicity of serotonin injected into them, whereas injection of cortisol into adrenalectomized mice restores their resistance to that amine (Higginbotham, 1962). The mechanism of this phenomenon has not yet been identified although it is claimed that the increase in sensitivity of adrenalectomized animals to serotonin is connected with a disturbance of its metabolism.

Adrenalectomy in rats leads to a twofold rise in the serotonin content in the peripheral tissues. Injection of cortisone into such animals restores its normal level (Hicks and West, 1958b). The blood serotonin level is lowered after adrenalectomy in rats or during traumatic shock (Medakovic, 1959), while its level in the liver is increased (Yeh et al., 1959).

Adrenalectomy is also accompanied by an increase in the excretion of serotonin metabolites in the urine (Enerbäck, 1960; Schwemmle et al., 1961; Put and Meduski, 1962), by a disturbance of the mechanisms of its binding in the body (Put and Meduski, 1962), and by a change in the serotonin concentration in the brain (De Maio and Marobbio, 1961; Pfeifer et al., 1963) and in the blood (Medakovic, 1959; Schmid et al., 1960a).

With regard to changes in the serotonin concentration in the brain after adrenalectomy, conflicting results have been obtained. According to Put and Meduski (1962) and to Pfeifer et al. (1963), adrenalectomy leads to a decrease in the serotonin concentration in the brain, which is evidently linked to some extent with a decrease in tryptophan hydroxylase activity (Azmitia and McEwen, 1966), while De Maio and Marobbio (1961), on the other hand, conclude that its concentration rises. De Maio (1959a) showed that 24 h after bilateral adrenalectomy the serotonin concentration in the blood serum falls while its concentration in the brain (cerebral hemispheres, base of the brain, medulla) rises. Similar changes (except in the cerebral hemispheres) also were observed after hypophysectomy. In both the first and second series of these experiments the greatest increase was observed in the medulla.

However, to evaluate the change in the serotonin level, De Maio (1959a) used a biological test based on the rat uterus, which also reacts to substance P, the concentration of which may rise during preparation of the brain extract (Towne and Scherman, 1960). If the serotonin concentration in the brain 24 h after bilateral adrenalectomy was investigated by a more accurate method (spectrofluorometry) (Towne and Scherman, 1960), no changes were found. Its contents in the central nervous system was unchanged not only after adrenalectomy, but also after hypophysectomy (Yeh et al., 1959; Resnik et al., 1961). These results were confirmed by more recent work (McKennee et al., 1965/1966): after adrenalectomy in rats the serotonin concentration was unchanged in the hypothalamus, in the pineal body, in the mesencephalon, and in the sensomotor cortex. Its concentration in the brain also was unchanged after administration of ACTH (Resnik et al., 1961).

Despite these conflicting results, some connection must be assumed between adrenocortical function and serotonin metabolism. This assumption is in harmony with the results of experiments in which adrenocortical hormones were administered.

Administration of Adrenocortical Hormones

Prolonged administration of adrenocortical hormones to patients with various diseases leads to an increase in the blood serotonin level (Schmid et al., 1960b), evidently as a result of its liberation from the tissue depots (Telford and West, 1960).

After administration of cortisone, the serotonin concentration in the rat liver falls (Yet et al., 1959). Similar results were obtained by investigation of the skin after administration of glucocorticoids (Hicks and West, 1958a).

Repeated (for 4-5 days) intramuscular injections of cortisone, prednisolone, triamcinolone, dexamethasone, and fludrocortisone lowered the serotonin concentration in the skin and small intestine of rats. Meanwhile, its concentration was increased in the gastric mucosa (Telford and West, 1960). In other experiments on rats

(De Maio and Marobbio, 1961), administration of prednisolone or DOCA lowered, while cortisone raised, the serotonin concentration in the brain. Cortisol restored the metabolism of this amine to its normal level in the serotonin-containing neurons of the rat brain (Fuxe et al., 1970), while corticosterone restored tryptophan hydroxylase activity when lowered by adrenalectomy (Azmitia and McEwen, 1969). Meanwhile, according to McKennee et al., (1965/1966), administration of hydrocortisone (0.75-5 mg/kg daily) to rats for 5 days did not change the level of this amine in the hypothalamus, pineal body, mesencephalon, and sensomotor cortex.

The conflicting results of these experiments naturally provide only indirect evidence of a link between serotonin and adrenocortical function, for adrenalectomy or repeated injection of adrenocortical hormones is itself accompanied by serious disturbances of many forms of metabolism. Secondary changes in serotonin metabolism may arise against this background.

Stress and the Serotonin Content in the Body

If serotonin is considered to have a role in the regulation of pituitary adrenal function, changes in its metabolism would be expected under extremal conditions. In fact, as several investigators have shown, the content of this amine does change after exposure to various stressors. For example, stress induced by a low temperature is accompanied by elevation of the blood serotonin level in rats (Gordon, 1961). Toh (1960) found an increase in the serotonin concentration in the brain of rats kept at a high or low temperature. Stress induced by injury to the hind limb in rats is accompanied by an increase in the serotonin concentration and in the rate of its metabolism in the brain stem (Stoner and Elson, 1971). An increase in the serotonin concentration (Goldberg and Salama, 1969) and in the rate of its metabolism (Bliss et al., 1968) were found in the brain of rats during stress induced by electric shock. An increase in the serotonin level was found both 10 min and 48 h after electric shock in the brain of rats, guinea pigs, rabbits, and dogs (Garattini and Valzelli, 1957). The greatest increase in guinea pigs was observed in the cerebral cortex (Poloni, 1957), while in rats and dogs it was found in the mesencephalon and diencephalon (Garattini et al., 1960). The observations of Fresia et al. (cited by Mantegazzini, 1966) showed that the greatest increase in serotonin concentration after electric shock in dogs occurs in the motor cortex (51%), hypothalamus (38%), and mesencephalon (60%).

Breitner et al. (1961) found an increase in the serotonin concentration in the brain stem of cats and rats after electric shock produced by general electrical stimulation. In the case of "local" electrical stimulation, on the other hand, when the electrodes were inserted above the sella turcica and into the occipital region, no changes in the serotonin concentration were found in the brain. In a later investigation (Breitner et al., 1964), however, in which a different strength of current was used for local electrical stimulation, a significant increase (by 14.4%) was found in the serotonin concentration in the brain stem of cats after 10 min, and the electrical stimulation was not accompanied by convulsion. Simultaneous estimation of noradrenalin showed no statistically significant change in its concentration in the part of the brain investigated. This increase in the serotonin level was evidently linked to some extent with an increase in its synthesis, because the formation of labeled serotonin from labeled trytophan injected previously is increased in the brain stem of rats subjected to electrical stimulation (Thierry et al., 1968). Meanwhile, some workers found no change in the serotonin concentration in the brain stem (Maynert and Levy, 1964) and in the blood (Green et al., 1957) during exposure to electric shock.

The parameters of the electrical stimulation applied are evidently an important factor determining the conflicting nature of these results. For instance, after electroconvulsive therapy the excretion of 5-hydroxyindoleacetic acid was considerably reduced. If, however, the intensity of electrical stimulation was increased, the excretion of this metabolic product of serotonin into the urine was greatly increased (Garattini and Valzelli, 1965).

The serotonin level was increased in the telencephalon, hippocampus, diencephalon, and mesencephalon of rabbits during insulin coma (Garattini and Valzelli, 1965). According to the same workers (Garattini and Valzelli, 1957) anoxia also leads to an increase in the serotonin concentration in the brain of animals. However, during stress induced by immobilization of an animal (Corrodi et al., 1968) or by horizontal oscillation (Rosecrans, 1969, 1970), the serotonin concentration in the brain falls.

Experiments on many species of laboratory animals have thus shown conclusively that stressors, which are known to stimulate the function of the pituitary-adrenal system, often produce accompanying changes in the serotonin concentration at the periphery and also in various parts of the brain, especially the brain stem. Serotonin can thus be confidently regarded as a biologically active substance participating in the adaptation syndrome (Miline et al., 1958).

Effect of Serotonin on the Pituitary-Adrenal Complex

Although the role of serotonin as a corticotrophin-releasing factor is rejected by many writers (Guillemin, 1957; Smelik and De Wied, 1958), many others have established its property of stimulating the pituitary-adrenocortical system.

Intraperitoneal injection of serotonin into mice (Halberg, 1954) and rats (Steiner and Hedinger, 1956) is accompanied by eosinopenia. The response of the eosinophils depends on the dose of serotonin injected; for example, a dose of 250 μmole/kg causes eosinopenia (Steiner and Hedinger, 1956), while a dose of 80 μmole/kg is accompanied by the development of eosinophilia (Steiner et al., 1957). Scaltrini (1956) found no change in the white blood cell count in persons receiving injections of serotonin (5–10 mg intramuscularly).

Some workers have concluded that serotonin exerts its influence on the pituitary-adrenal system through the pituitary. This conclusion is based on the fact that injection of serotonin into rats causes a decrease in the ascorbic acid concentration in the adrenals (Bertelli et al., 1954; Georges, 1957; Moussatche and Alvares-Pereiro, 1957; Fischer et al., 1959; Fiore-Donatti et al., 1959), whereas in hypophysectomized rats the ascorbic acid concentration is unchanged (Moussatche and Alvares-Pereiro, 1957; Miyawaki et al., 1961).

There is no general agreement regarding the possibility of a direct effect of serotonin on the adrenal cortex in the literature. During perfusion of the isolated adrenals of dogs, serotonin was found to stimulate the secretion of hydrocortisone by almost the same degree as after injection of ACTH (Verdesca et al., 1961). The activating effect of serotonin on corticosterone secretion by cells of the adrenal cortex in rats has also been demonstrated by

Haning et al. (1970) in experiments *in vitro*. However, other workers who added serotonin to the incubation medium found that it did not stimulate corticosterone production by adrenal tissue taken from intact rats (Jouan, 1967; Müller and Huber, 1969). The absence of an activating effect of serotonin on the adrenal cortex has also been observed in experiments in which the serotonin precursor, 5-hydroxytryptophan, was injected into hypophysectomized rats (Popova et al., 1972b).

Paradoxical as it may seem, injection of serotonin (in the way in which this was done by the authors cited above, i.e., by the usual methods of subcutaneous, intramuscular, intraperitoneal, and intravenous injection) is by no means the best method of studying its physiological effects. The reasons are connected with some peculiarities of this biogenic amine: the rapidity with which it is destroyed, its inability to penetrate through the blood–brain barrier, which is particularly important where detection of the central effects are concerned, and so on, and it will be dealt with again below. Because of these properties of serotonin, when the role of serotoninergic structures is studied, attempts must be made to change its tissue level by action in other ways.

Disturbance of the Serotonin Balance and the Pituitary-Adrenal Complex

The role of serotonin in the activity of the pituitary-adrenal system can be studied by several methods which lead to disturbance of the dynamic equilibrium of endogenous serotonin. The balance of biogenic amines in the body depends, first, on the intensity of their formation from precursors; second, on the ability of the tissues to retain biogenic amines; and third, on the rate of their destruction. By acting on any one of these factors it is possible to reproduce changes in the tissue serotonin concentration.

Changes in the Rate of Destruction

Monoamine Oxidase Inhibitors and Serotonin Metabolism. Monoamine oxidase is now considered to be the chief enzyme decomposing serotonin to 5-hydroxyindoleacetaldehyde, which is subsequently oxidized to 5-hydroxyindoleacetic acid, the end product of serotonin metabolism (Titus and Udenfriend, 1954; Udenfriend, 1958). Other pathways of serotonin breakdown have also been found:

by *n*-acetylation (McIsaac and Page, 1959) and by binding of the hydroxyl group with glucuronic acid (Weissbach et al., 1961), although their relative importance is slight and 5-hydroxyindoleacetic acid, the result of oxidative deamination, accounts for more than 90% of the total excretion of the 5-hydroxyindoles (Weissbach et al., 1961).

This principal pathway of serotonin breakdown can be blocked effectively by administration of monoamine oxidase inhibitors. Compounds exist which inhibit monoamine oxidase; the most widely studied and extensively used of these substances are hydrazine derivatives.

The ability of isonicotinic acid isopropylhydrazide (iproniazid) to inhibit monoamine oxidase was first established in Zeller's laboratory (Zeller and Barsky, 1952; Zeller et al., 1952a,b) and subsequently confirmed repeatedly. Administration of iproniazid quickly produced prolonged inhibition of monoamine oxidase both *in vitro* (Zeller et al., 1955; Spector et al., 1960) and *in vivo* (Gey and Pletscher, 1961). Meanwhile, the original member of the group of hydrazine derivatives, isonicotinic acid hydrazide (isoniazid), hardly inhibited monoamine oxidase at all (Zeller et al., 1952a,b). Administration of monoamine oxidase inhibitors leads to an increase in the tissue serotonin level (Udenfriend et al., 1957a; Shore, 1958; Brodie et al., 1959), especially in the brain.

Monoamine oxidase catalyzes the oxidative deamination of all monoamines, including the active ones such as the catecholamines. After injection of monoamine oxidase inhibitors, both the serotonin level and the tissue catecholamine level rise, although these changes in the various levels differ in degree and take place at different rates. The increase in the concentrations of serotonin, noradrenalin, and dopamine in the brain depends on the properties of the inhibitor and the dose used. However, the level of these biologically active substances after administration of a particular monoamine oxidase inhibitor rises by an unequal degree in different parts of the brain (Valzelli and Garattini, 1968).

The reason is that, by contrast with serotonin, the principal pathway of breakdown of noradrenalin and adrenalin is by O-methylation, catalyzed by the enzyme catechol-O-methyltransferese, while oxidative deamination plays a relatively unimportant role

(Axelrod, 1957, 1959; Axelrod and Tomchick, 1958; Axelrod et al., 1959). Meanwhile, the tissue noradrenalin is much less sensitive to the action of monoamine oxidase inhibitors than serotonin, which lacks a hydroxyl group and is a far better substrate for monoamine oxidase (Kopin, 1964; Gatgounis, 1965). After administration of monoamine oxidase inhibitors the serotonin concentration thus rises sooner and to a greater degree than the noradrenalin concentration (Brodie et al., 1959; Spector et al., 1960; Leroy and Van der Schoot, 1962). This difference is the basis for the use of monoamine oxidase inhibitors in the analysis of the action of serotonin on the pituitary-adrenal system.

Monoamine Oxidase Inhibitors and Adrenocortical Function. The facts in the literature relating to the effect of monoamine oxidase inhibitors on adrenocortical function are few in number and, more important, conflicting in nature. Sapeika (1959) states that iproniazid, in a dose of 150-300 mg/kg, stimulates adrenocortical function in rats. Similar results have been obtained in man during prolonged administration of the compound (Crane and Wolfman, 1960).

On the other hand, determination of the corticosterone concentration in the blood and adrenals failed to reveal any stimulant action of iproniazid on adrenocortical function after a single injection into rats in a dose of 100 mg/kg (Eechaute et al., 1962a,b). In the opinion of Kivalo and Rinne (1961), this inhibitor does not stimulate the anterior lobe of the pituitary. Finally, some workers have described the inhibitory action of iproniazid on the hormonal activity of the adrenal cortex in mice (De Schaepdryver and Preziosi, 1959) and rats (Georges and Herold, 1958).

The work of De Maio (1959b) showed that phenelzine, another monoamine oxidase inhibitor, causes a decrease in the ascorbic acid concentration in the adrenals of rats. Administration of tranylcypromine increases the blood level of corticosterone simultaneously with an increase in the brain serotonin concentration (Dixit and Buckley, 1969).

Because of these conflicting results, it was decided to carry out experiments with iproniazid (Naumenko and Popova, 1965) in order to compare its action on the adrenal cortex with that of isoniazid, which has virtually no inhibitory action on monoamine oxidase (Zeller et al., 1952a,b).

TABLE 10. 17-Hydroxycorticosteroid Concentration in Blood of Guinea Pigs after Administration of Monoamine Oxidase Inhibitors

Compound	Mean plasma corticosteroid level ±m, μg/100 ml			Number of animals
	Initial	After administration of hydrazides		
		1 h	4 h	
Iproniazid, 50 mg/kg	26.01 ±2.09	35.6 ±3.08 (P < 0.02)	69.0 ±4.92 (P < 0.001)	18
Isoniazid, 65 mg/kg	34.6 ±3.09	69.2 ±4.32 (P < 0.001)	77.6 ±7.53 (P > 0.1)	18

The results of these experiments showed that both substances increase the 17-hydroxycorticosteroid concentration in the peripheral blood plasma of guinea pigs (Table 10). Nevertheless, characteristic differences were found in the response of the adrenals to injection of iproniazid and isoniazid.

A significant increase in the corticosteroid concentration in the peripheral blood plasma was observed 1 h after injection of both compounds, but isoniazid led to a considerably greater increase than iproniazid (by 100 and 40% respectively). Whereas 4 h after injection of isoniazid the corticosteroid level remained the same, iproniazid continued to stimulate adrenocortical function. Other results showing excitation of the activity of these endocrine glands by iproniazid have been obtained by Zoryan (1965).

An increase in the corticosteroid level after administration of isoniazid was observed in experiments on rats (Simila and Laine, 1963) and after its prolonged administration to patients (Rausch-Stroomann and Witthöft, 1956; Romero and Solohaga, 1961).

Differences in the response of the adrenal cortex to iproniazid and isoniazid in our experiments suggest that for iproniazid there are other mechanisms of stimulation of adrenocortical function. The fact that iproniazid inhibits monoamine oxidase more strongly than isoniazid and contributes to the accumulation of biogenic amines suggests that the response of the adrenal cortex 4 h after its administration can be attributed to this property. In the earlier stages iproniazid, like isoniazid, increases the corticosteroid concentration in the peripheral blood evidently by different mechanisms, unconnected with monoamine oxidase inhibition.

As has been stated already, after inhibition of monoamine oxidase the serotonin level rises first, and the tissue noradrenalin concentration only begins to rise much later. As Brodie et al. (1959) found, the brain serotonin level in rabbits doubles during the first 6 h after injection of iproniazid (100 mg/kg), whereas the noradrenalin concentration in the brain is still almost unchanged at this time. Its level begins to rise later. The experiments described above with iproniazid accordingly suggest that the increase in intensity of stimulation of the pituitary-adrenal system takes place as the result of an increase in the serotonin level in the body.

Besides the use of monoamine oxidase inhibitors, there is another way in which the problem of immediate concern to us can be studied, namely, by examining the effect of reserpine on the ability of the tissues to retain serotonin.

Disturbance of the Ability of the Tissues to Retain Serotonin

Despite the use of different methods to assess pituitary-adrenal function, many investigators have shown that reserpine stimulates adrenocortical function in animals of different species and in man (Guillemin, 1957; Harwood and Mason, 1957; Girod, 1962; Eechaute et al., 1962a,b; Westermann et al., 1962; Newcomer, 1962; Westermann, 1965; Feldman et al., 1965/1966; Giuliani et al., 1966). The activating effect of reserpine is absent after hypophysectomy (Westermann, 1965).

My own experiments confirmed that reserpine stimulates pituitary-adrenal function in guinea pigs if a single dose of 5 mg/kg of this substance is administered to them by mouth (Table 11).

As Table 11 shows, the stimulating effect of reserpine is prolonged: not until 72 h after its administration did the blood corticosteroid level return to its initial value. This indicates that reserpine does not act directly, but through exhaustion of the monoamine depots,* for its elimination from the body continues for several hours (Hess et al., 1956). However, reserpine is known

*There is an alternative view on the mechanism of action of reserpine to explain its stimulant effect on the pituitary-adrenal system: it is claimed that since reserpine prevents the binding of serotonin, its action is accompanied by the constant liberation of very small concentrations of free serotonin into the blood stream (Holtz et al., 1957; Brodie et al., 1959).

TABLE 11. Blood 17-Hydroxycorticosteroid Concentration at Different Times after Administration of Reserpine to Guinea Pigs

Time of investigation	Mean plasma corticosteroid level ±m, μg/100 ml		P	Number of animals
	Initial	Reserpine		
After 24 h	35.0±8.64	103.5±4.70	<0.001	10
» 48 h	40.9±3.37	82.5±3.32	<0.001	10
» 72 h	46.1±4.39	51.3±4.35	>0.1	10

to act not only on the serotonin level but also on the concentration of catecholamines in the body, not only in the central nervous system but also at the periphery (Holzbauer and Vogt, 1954; Carlsson et al., 1957; Waund et al., 1958; Paasonen and Krayer, 1959; Malhotra and Prasad, 1962). For this reason, when using reserpine, it is difficult to differentiate between effects due to its action on each monoamine separately and to its action on the pituitary-adrenal system. However, there is evidence to suggest that the stimulant effect of reserpine is due principally to its influence on the serotonin of the brain. This problem has been specially discussed by Westermann (Westermann et al., 1962; Westermann, 1965).

Stimulation of pituitary-adrenal function has been shown (Westermann et al., 1962) to take place not on account of the peripheral catecholamines, for 1) an increase in adrenocortical function was observed after injection of reserpine into demullated rats; 2) syrosingopine, an analogue of reserpine, which in small doses causes the liberation of noradrenalin from peripheral nerve endings, did not stimulate adrenocortical function. Only large doses of syrosingopine, which at the same time lowered the monoamine concentration in the brain, possessed this action.

On the other hand, isoreserpine and isoraunescine, isomers of reserpine and raunescine, did not stimulate the pituitary-adrenal system and did not alter the monoamine concentration in the brain. Only in a near-lethal dose (50 mg/kg) did isoreserpine stimulate adrenocortical function very slightly.

It can be concluded from these experiments that the action of reserpine is most probably linked with the mobilization of the brain serotonin. Evidence in support of this view is given by experiments in which α-methyl-m-tyrosine was given, for this compound in a dose of 400 mg/kg depleted the catecholamine reserves of the brain for more than 24 h without any significant changes in its serotonin concentration. Administration of this amino acid did not stimulate adrenocortical function. At the same time, injection of reserpine 24 h after α-methyl-*m*-tyrosine led to definite activation of the pituitary-adrenal system (Westermann et al., 1962; Westermann, 1965).

The results of the experiments with reserpine thus suggest that the serotonin of the central nervous system activates the pituitary-adrenocortical system, and they provide an incentive for the search for further evidence.

Changes in the Intensity of Synthesis

Relatively recent investigations have shown that two specific enzyme systems in the animal body participate in the synthesis of serotonin (Clark et al., 1954; Udenfriend et al., 1957a,b; Bogdanski et al., 1958a,b; Udenfriend, 1958): 1) tryptophan oxidase, oxidizing L-tryptophan to 5-hydroxytryptophan; 2) tryptophan decarboxylase, converting 5-hydroxytryptophan into serotonin (Fig. 17).

The tissue serotonin concentration can be changed theoretically in two ways: by blocking serotonin synthesis and thereby de-

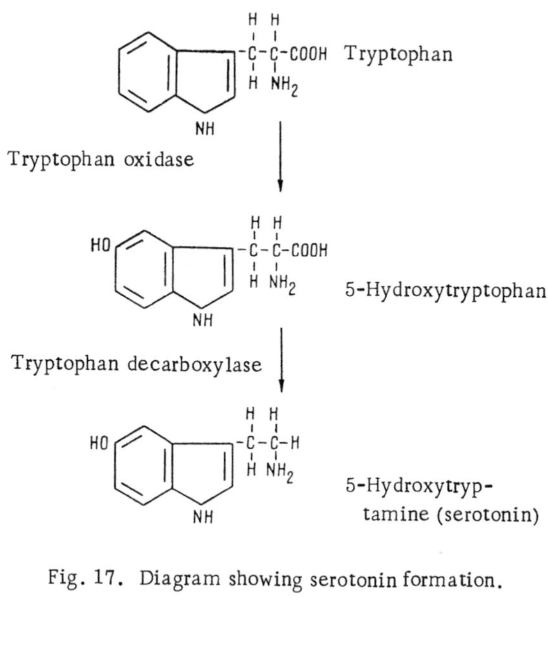

Fig. 17. Diagram showing serotonin formation.

creasing the tissue concentration of this amine, or by increasing the tissue serotonin concentration. In practice the first method was not available for researchers because no substances capable of blocking serotonin synthesis reliably and selectively in the body existed.

It is only in the last few years that a compound (*p*-chlorophenylalanine) has been discovered which, if injected subcutaneously into rats, causes their brain serotonin concentration to fall after 3 days by more than 90% of its initial value (Koe and Weissman, 1966). These workers found that *p*-chlorophenylalanine, in experiments *in vitro* and *in vivo*, blocked the tryptophan oxidase of the liver and prevented the accumulation of serotonin and 5-hydroxyindoleacetic acid in the brain after administration of tryptophan, but had only a slight effect on the increase in serotonin after administration of 5-hydroxytryptophan. They concluded from their experiments that *p*-chlorophenylalanine depletes the serotonin depots by blocking the enzyme responsible for the hydroxylation of tryptophan and its conversion into 5-hydroxytryptophan, the immediate precursor of serotonin. Moreover, *p*-chlorophenylalanine depleted the serotonin depots of the brain three times more intensively than at the periphery and had little or no effect on the tissue catecholamine concentration.

Lowering the Tissue Serotonin Concentration. The writer has used *p*-chlorophenylalanine in an attempt to study the role of serotonin in pituitary-adrenal function. The experiments were carried out at the Institute of Experimental Medicine of the Hungarian Academy of Sciences, in conjuction with E. Stark, K. Pfeifer, and J. Asch.

As preliminary experiments showed, *p*-chlorophenylalanine when injected subcutaneously into rats had a nonspecific action after 1 h on the pituitary-adrenal system, causing an increase of 263% in the blood corticosterone concentration. This was evidently connected with its stimulant action. The blood corticosterone level returned to its initial value 24 h after injection of the compound.

Three days after its administration, the blood corticosterone concentration of the rats was normal, while the brain serotonin level was lowered. The decrease in the brain serotonin concentration varied considerably. In different experiments it ranged between 51 and 20% of the initial concentration, and the decrease did not exceed 80% of any of the experiments. This pattern agrees with results obtained by other workers (Thierry et al., 1968).

Unfortunately, the marked variability and the insufficient degree of lowering of the brain serotonin concentration in the individual experiments prevented any precise interpretation of the results. Nevertheless, the lowering of the serotonin concentration by *p*-chlorophenylalanine was accompanied, during exposure to stress (formalin and naphazoline were used as stressors), by a tendency toward inhibition of pituitary-adrenal function. The results suggest that the decrease in serotonin concentration, which evidently occurred first in the brain, has some effect on the functional state of this system under the conditions of stress, while experiments in which a disturbance of the diurnal fluctuations in the blood corticosteroid and brain serotonin levels was found after administration of *p*-chlorophenylalanine are evidence that serotonin may play a role in circadian rhythms (Scapagnini et al., 1971). This, in turn, suggests that the serotonin of the brain may play a part in the regulation of pituitary-adrenal function. This hypothesis is in agreement with the results of experiments in which monoamine oxidase inhibitors and reserpine were used. However, it must be borne in mind that the use of these substances does not guarantee selectivity in the modification of serotonin metabolism only. Allowance must be made for the fact that the substances used also influence catecholamine metabolism. This applies also to *p*-chlorophenylalanine, for in contrast with the results obtained by Koe and Weissman (1966), more recently reports have been published on the marked effect of this substance on the catecholamine concentration in the brain (De Schaepdryver et al., 1969; Keller, 1972). The hypothesis of the possible role of the brain serotonin as a regulator of pituitary-adrenal function thus requires more direct evidence.

Elevation of the Tissue Serotonin Concentration. The most direct method of affecting the level of the biogenic amines is by the use of the precursors of their biosynthesis.

Administration of serotonin by the usual methods of injection in order to analyze serotoninergic structures is less demonstrative in its results than administration of its precursor: 5-hydroxytryptophan leads to the accumulation of serotonin in places where it is formed naturally, including in the brain. Furthermore, administration of 5-hydroxytryptophan enables the tissue serotonin level to be maintained comparatively high for a longer time, whereas serotonin injected parenterally is rapidly destroyed (Udenfriend et al., 1957a,b; Bogdanski et al., 1958a; Costa, 1960).

Consequently, the administration of 5-hydroxytryptophan reveals the physiological properties of serotonin much better than serotonin itself if administered by the usual routes (Bogdanski et al., 1958a,b). At the same time, there are reports in the literature that in man (Schwemmle et al., 1961) and in experiments on the isolated dog adrenal (Verdesca et al., 1961) 5-hydroxytryptophan did not stimulate the adrenal cortex. It was therefore decided to investigate whether the effect of administration of exogenous serotonin was similar to that of administration of exogenous 5-hydroxytryptophan (Naumenko, 1965d). Experiments showed that serotonin and its precursor, 5-hydroxytryptophan, stimulate the pituitary-adrenal system (Table 12).

The absence of any stimulant action of the serotonin precursor on adrenocortical function in patients (Schwemmle et al., 1961) is probably due to the use of very small doses (0.2-1 mg/kg). To judge from the experiments of Verdesca et al. (1961), 5-hydroxytryptophan has no direct stimulant effect on steroid production in the adrenals. Presumably, therefore, the action of the precursor is the result of its conversion into serotonin, especially in the brain, and the subsequent effect of this amine on central serotoninergic structures. In the experiments described above any possible effect of 5-hydroxytryptophan through a decrease in the noradrena-

TABLE 12. Blood 17-Hydroxycorticosteroid Concentration in Guinea Pigs 1 h after Injection of Serotonin and 2.5 h after Injection of 5-Hydroxytryptophan

Substance injected	Mean plasma corticosteroid concentration ± m, μg/100 ml	P	Number of animals
Distilled water	42.6 ±2.87	< 0.001	20
Serotonin, 2 mg/kg. .	148.7 ±10.28		
Distilled water. . . .	50.8 ±4.76		
5-Hydroxytryptophan, 100 mg/kg.	121.2 ±4.47	< 0.001	15

TABLE 13. Blood 17-Hydroxycorticosteroid and Brain Serotonin Concentrations in Guinea Pigs 2.5 h after Subcutaneous Injection of 5-Hydroxytryptophan

Dose of 5-hydroxytryptophan, mg/kg	Mean plasma corticosteroid concentration ± m, μg/100 ml	Mean brain serotonin concentration ±m, μg/g	Number of animals
Not given	25.0 ±6.11	0.263 ±0.010	3
25	28.0	0.303	1
50	49.0 ±2.51	0.443 ±0.126	3
100	100.0 ±4.39	0.680 ±0.074	4

lin concentration in the brain was ruled out because it does not possess this action even in much larger doses (Brodie et al., 1966).

The dependence of the stimulant action of 5-hydroxytryptophan on pituitary-adrenal function on the brain serotonin level is confirmed by experiments in which the effects of different doses of this precursor on adrenocortical function were compared with the brain serotonin concentration. These experiments showed correlation between elevation of the brain serotonin level produced by 5-hydroxytryptophan and the accompanying stimulation of pituitary-adrenal function (Table 13).

The results described above collectively provided weighty evidence for regarding the brain serotonin as a substance capable of participating in the regulation of the pituitary-adrenal system. However, more direct evidence in support of this conclusion was naturally required. In the next experiments the method of injecting serotonin into the lateral ventricles was accordingly used (Naumenko, 1965c, 1966).

TABLE 14. Blood 17-Hydroxycorticosteroid Concentration in Guinea Pigs after Injection of Serotonin into the Lateral Ventricle

Substance injected	Mean plasma corticosteroid concentration ±m, μg/100 ml	P	Number of animals
Tyrode solution	59.3±7.21	>0.1	7
Serotonin, 50 μg	58.0±7.83		
Tyrode solution.	78.0±7.86	<0.05	11
Serotonin, 100 μg	102.0±6.34		
Tyrode solution	69.0±5.49	<0.001	28
Serotonin, 200 μg	127.7±9.51		

The Central Action of Serotonin

Intraventricular Injection of Serotonin into Animals with an Intact Brain

Injection of serotonin through a cannula inserted into the lateral ventricle was accompanied by an increase in the 17-hydroxycorticosteroid concentration in the peripheral blood plasma of guinea pigs. This increase was proportional to the dose injected (Table 14).

The results of these experiments show that injection of serotonin in a dose of 50 μg did not stimulate the pituitary-adrenal system. Larger doses (100 and 200 μg) led to increase in the blood corticosteroid concentration of the guinea pigs on the average by 53 and 85% respectively. Since serotonin creatinine-sulfate was used in these experiments, doses of this salt of 50, 100, and 200 μg were equivalent to actual doses of 21.5, 43, and 86 μg serotonin respectively (Garattini and Valzelli, 1965).

These experiments clearly demonstrate the stimulant action of serotonin on the pituitary-adrenal system through excitation of central serotoninergic structures. However, experiments in which serotonin is injected into the cerebral ventricles of animals with an intact brain cannot completely rule out the possibility of excitation of corresponding structures at the periphery followed by a secondary response of the pituitary-adrenal system. After stimulation of the central serotoninergic structures, nervous impulses arising in the brain may produce effects at the periphery by transmission along efferent pathways through the spinal cord. In order to differentiate between the direct central action of this amine and its possible indirect effects, in the next experiments (Naumenko, 1965c, 1966; Naumenko and Il'yuchenok, 1967) serotonin was injected into the lateral ventricle after mesencephalic brain section.

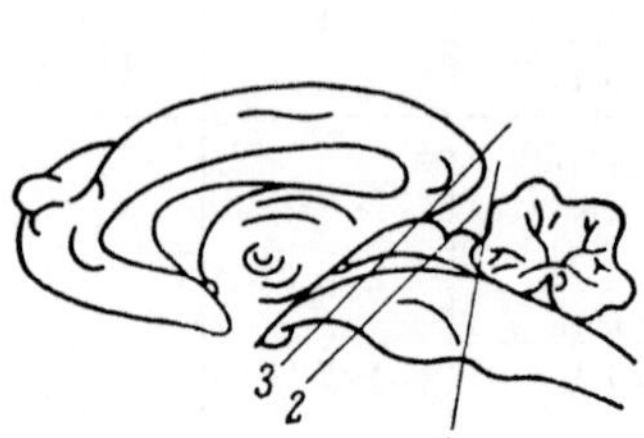

Fig. 18. Diagram showing planes of brain section. Explanation in text.

Injection of Serotonin into the Lateral Ventricle after Mesencephalic Section

The effect of serotonin on the hypothalamo-hypophyseo-adrenal system was studied after the following types of brain section (Fig. 18): 1) caudally to the inferior colliculi and posteriorly to the pons; 2) between the superior and inferior colliculi or through the anterior part of the inferior colliculi and anteriorly to the pons; 3) through the superior colliculi and between the mammillary bodies of the hypothalamus and the pons. The plane of section thus passed caudally to the site of insertion of the cannula.

One hour after injection of serotonin into the lateral ventricle of the animals with brain section the 17-hydroxycorticosteroid concentration in the peripheral blood plasma was raised (Table 15). However, the difference in the degree of elevation of the corticosteroid level in response to the serotonin injection in the animals with the various types of brain section was not statistically significant.

When it became impossible for serotonin to act indirectly through peripheral mechanisms, for the brain section blocked the transmission of impulses along descending tracts, this amine thus continued to stimulate the hypothalamo-hypophyseo-adrenal system.

Consequently, excitation of central serotoninergic structures has a direct stimulant effect on the hypothalamo-hypophyseo-adrenocortical system.

TABLE 15. Blood 17-Hydroxycorticosteroid Concentration after Injection of Serotonin into Lateral Ventricle of Guinea Pigs with Mesencephalic Section

Substance injected	Level of section*	Mean plasma corticosteroid level ±m, μg/100 ml		P	Number of animals
		Initial	After injection of substance		
Tyrode solution	1	65.5 ±3.95	68.6 ±5.33	> 0.1	11
	2	69.9 ±8.77	71.8 ±8.49	> 0.1	6
	3	66.8 ±6.36	74.3 ±6.80	> 0.1	5
Creatine-sulfate	1	54.8 ±5.91	51.2 ±4.42	> 0.1	8
	2	61.7 ±3.57	51.9 ±6.64	> 0.1	5
Serotonin, 200 μg	1	63.1 ±3.60	92.5 ±6.15	< 0.001	20
	2	72.3 ±5.10	114.4 ±8.97	< 0.001	20
	3	67.9 ±6.74	112.1 ±10.68	< 0.001	12

*Explanation in text.

Localization of Serotoninergic Structures in the Brain

It was shown in the preceding section that serotonin has a direct central action on the pituitary-adrenocortical system. I consider that this is a fact of fundamental importance because it shows that the serotoninergic structures of the brain participate directly in the mechanisms controlling the adrenocorticotrophic function of the anterior pituitary.

However, the results of experiments in which this biogenic amine was injected into the lateral ventricle cannot of course shed light on the localization of those serotoninergic structures whose excitation influences pituitary-adrenal activity. The method used to investigate this problem was by chronic implantation of microcannulas through which serotonin was applied directly to particular anatomical structures of the brain.

Determination of the precise topography of the serotoninergic structures over the whole central nervous system is, of course, an impossible task for one investigator. I therefore studied their localization only in the hypothalamus, the mesencephalon, and limbic structures of the telencephalon (Naumenko, 1967b, 1968; Naumenko et al., 1968).

Hypothalamus

Local Injection of Serotonin into Animals with an Intact Brain. After local injection of serotonin, no marked changes were found in the animals' behavior apart from some limitation of their movements. As a rule, the guinea pigs sat quietly in one place for most of the time.

To study the role of serotoninergic structures of the hypothalamus, microcannulas were inserted into the posterior, central, anterior, and lateral zones of the hypothalamus, into the preoptic region, and into the subthalamus (Fig. 19).

One hour after local injection of serotonin into the medial mammillary nucleus and supramammillary region, the posterior hypothalamic nucleus, the dorso- and ventromedial nuclei of the central hypothalamus, the anterior hypothalamus, preoptic region, and subthalamus of the guinea pigs, a clear increase in the 17-hydroxycorticosteroid concentration in the peripheral blood was observed. Meanwhile, injection of serotonin into the lateral mam-

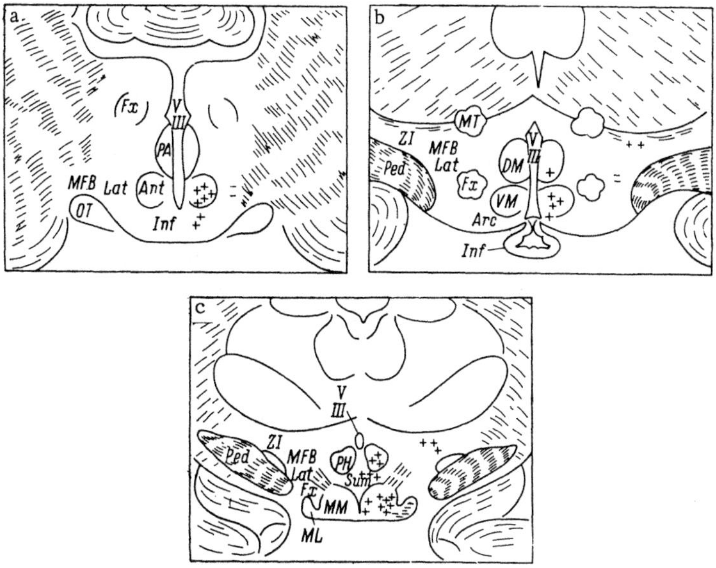

Fig. 19. Diagram of anterior (a), central (b), and posterior (c) zones of the hypothalamic region. Stimulant effect of serotonin (+). No response to serotonin (-). Ex) Fornix; V III) third ventricle; PA) paraventricular nucleus; Ant) anterior hypothalamic nucleus; Lat) lateral hypothalamus; Inf) region of infundibulum; OT) optic tract; MFB) medial forebrain bundle; ZI) subthalamus; Ped) cerebral peduncles; DM) dorsomedial nucleus; VM) ventromedial nucleus; Arc) arcuate nucleus; PH) posterior hypothalamic nucleus; Sum) supramammillary nucleus; ML) lateral mammillary nucleus; MT) mammillothalamic tract; MM) medial mammillary nucleus.

TABLE 16. Blood 17-Hydroxycorticosteroid Concentration in Guinea Pigs after Local Injection of Serotonin into the Hypothalamus

Localization of cannula*	Mean plasma corticosteroid level ±m, μg/100 ml		P	Number of animals
	Physiological saline	Serotonin†		
MM	43.5 ±2.48	84.3 ±9.57	< 0.002	8
Sum	34.5 ±2.62	92.3 ±11.23	< 0.01	4
PH	51.9 ±7.09	114.8 ±21.38	< 0.05	3
ML	57.5 ±8.91	58.0 ±9.85	> 0.1	6
VM	39.7 ±8.53	101.5 ±14.49	< 0.02	4
DM	45.1	77.6		1
Ant	51.6 ±5.74	124.3 ±14.17	< 0.001	7
RPO	56.5 ±4.55	140.5 ±9.26	< 0.001	8
Lat	49.0 ±5.93	48.3 ±3.14	> 0.1	4
ZI	33.9 ±3.86	73.0 ±8.40	< 0.002	5

*Explanation of abbreviations in caption to Fig. 19. RPO) Preoptic region.

†Here and in subsequent experiments 10 mg serotonin creatinine-sulfate, corresponding to 4.3 μg serotonin, was injected.

millary nucleus and into the lateral hypothalamus was not followed by stimulation of pituitary-adrenal function (Table 16).

The results of these experiments thus show that serotonin, when injected into certain nuclear groups in various zones of the hypothalamic region, stimulates the function of the pituitary-adrenal complex. However, the function of the pituitary-adrenal system is not stimulated in every case, but only if the serotonin is injected into nuclei lying closer to the medial plane of the hypothalamus and into the subthalamus. Its injection into the lateral zones of the hypothalamic region did not change the activity of the pituitary-adrenocortical system, whether the injection was made into zones of the anterolateral and lateral hypothalamus or the posterolateral hypothalamus. It can be concluded from these results that serotoninergic structures related to pituitary-adrenal function are located in the medial zones of the hypothalamic region throughout its extent and in the subthalamus.

The degree to which the blood corticosteriod level was raised in these experiments after injection of serotonin into the various hypothalamic zones was approximately the same: by 2-2.5 times. This result suggests that the intensity of distribution of serotoninergic structures is the same in different parts of the hypothalamus. This assumption agrees with experiments in which the serotonin concentration was determined in the anterior and posterior hypothalamus in animals and man. Their results showed that the serotonin concentration was the same in these parts. The facts described above are summarized in Table 17, which was compiled by Garattini and Valzelli (1965) from results obtained by various workers.

The results of my experiments agree with those of other investigations giving, according to Bloom et al. (1963), direct proof of the existence of serotoninergic structures in the hypothalamic region. Bloom found that neurons responding to microelectrophoretic injection of serotonin, unlike neurons responding to noradrenalin and acetylcholine, are distributed diffusely throughout the hypothalamus, from the mammillary bodies to the preoptic region.

From earlier observations it seems more likely that the morphological basis of the mechanism regulating ACTH secretion in the hypothalamus consists of diffuse nuclei or a nerve net than of a discrete cellular center (Moll, 1959, 1960; Brodish, 1963,

1964a,b). Stimulation of the pituitary-adrenal system was observed in my experiments after excitation of serotoninergic structures of the principal nuclear formations located in median areas of the anterior, central, and posterior hypothalamus. This extensive distribution of the zone responding to serotonin in the hypothalamus has been confirmed by other investigations (Bloom et al., 1963), and it is suggested that it contains a polysynaptic nerve net which is serotoninergic, at least partly, in its chemical nature.

The results correspond to those obtained by many other investigators who have studied the role of hypothalamic nuclei in the regulation of the pituitary-adrenocortical system by the use of an electrical stimulation method. Experiments showing that electrical stimulation of the hypothalamus by means of electrodes implanted medially between the optic chiasma and the mammillary bodies, including the latter structures, is accompanied in many species of animals by stimulation of the pituitary-adrenal system, were described in Chapter I. Meanwhile, stimulation of the lateral zones of the hypothalamus does not change the pituitary-adrenal function.

These results and those of my own experiments suggest that the serotoninergic structures of the median zones of the hypothalamus participate in the mechanism of the response of the pituitary-adrenal system to electrical stimulation. This hypothesis agrees almost completely with results obtained by H. Krieger and D. Krieger (1970) and with the experiments in which electrical stimulation of the anterior and central hypothalamus caused an increase in the serotonin concentration in blood flowing from the brain, on the basis of which it was concluded that these zones contain serotoninergic structures (Gromova, 1966; Vekshina, 1967). Meanwhile, there is only partial agreement between my own results and those of Shreiberg (Shreiberg and Dunaeva, 1965; Shreiberg, 1966;

TABLE 17. Serotonin Concentration in Anterior and Posterior Hypothalamus of Mammals (Garattini and Valzelli, 1965), μg/g Brain Tissue

Zone of hypothalamus	Cat	Dog	Man
Anterior	0.24	0.22	0.35
Posterior	0.25	0.225	0.30

Shreiberg and Matlina, 1967), who observed stimulation of the adrenals in rats after injection of serotonin only into the central hypothalamus.

According to the workers described above, the anterior and posterior hypothalamus contain serotoninergic structures whose excitation is accompanied by inhibition of pituitary-adrenal function. However, they reached this conclusion after experiments in which 2.5-10 μg serotonin in agar-agar was injected into the hypothalamus 7-12 days before investigation of adrenocortical function. Now free serotonin is known to be unstable and to be constantly broken down by monoamine oxidase, the concentration of which in the hypothalamus is high. The synthesis and breakdown of serotonin are known to take place very rapidly in the brain stem (Udenfriend and Weissbach, 1968, cited by Garattini and Valzelli, 1965): the half-breakdown time of serotonin in the brain is 10-30 min. Garattini and Valzelli (1965) calculate that up to 150μg serotonin/kg body weight can be converted daily in the rat brain.

Considering this high intensity of serotonin metabolism, it is difficult to connect the changes in adrenocortical function observed by Shreiberg and co-workers with the action of 2.5-10 μg serotonin injected 1-1.5 weeks before the experiment (the quantities of serotonin in agar remaining in the hypothalamus at the time when adrenocortical function was investigated were not determined by these workers).

TABLE 18. Blood 17-Hydroxycorticosteroid Concentration after Local Injection of Serotonin into Hypothalamus of Guinea Pigs after Mesencephalic Section

Localization of cannula*	Mean plasma corticosteroid concentration ±m, μg/100 ml		P	Number of animals
	Initial	Serotonin		
PH	86.8 ±13.59	80.6 ±13.18	> 0.1	5
Sum	56.2	35.1		1
MM	77.7 ±6.76	117.4 ±11.35	< 0.05	5
{	78.4	118.2		2
VM {	76.5	154.0		
Ant	95.4 ±6.48	142.4 ±13.20	< 0.01	8
RPO	85.3 ±6.92	131.0 ±13.12	< 0.05	4

*PH) Posterior hypothalamic nucleus; Sum) supramammillary region; MM) medial mammillary nucleus; VM) ventromedial nucleus; Ant) anterior hypothalamic nucleus; RPO) preoptic region.

The discovery of each new fact brings in train a number of new problems. In this particular case the question of to which system of nerve fibers the serotoninergic structures of the hypothalamus belong arises. The possibility cannot be ruled out that the serotoninergic structures found in the hypothalamus are connected with endings of axons entering the medial zones of the hypothalamus in large numbers from the medial forebrain bundle. This bundle, the chief associative bilateral hypothalamic tract, runs from the telencephalon through the lateral zones of the preoptic region and hypothalamus to the mesencephalon. By means of the numerous nerve fibers which it gives off in a medial direction it connects the various zones and nuclei of the hypothalamus (Gloor, 1956). Injection of serotonin into the lateral zones of the hypothalamus did not change the function of the pituitary-adrenal system, so that the fibers themselves evidently do not respond to serotonin. Meanwhile, their endings and structures innervated by them are serotoninergic, and this is why in the experiments described above the blood corticosteroid level rose after injection of serotonin into the more medial areas of the hypothalamus.

According to Heller and Moore (1965), the medial forebrain bundle contains serotoninergic fibers. This is shown by the following facts: 1) after division of this tract in the region of the lateral hypothalamus in rats there is a decrease in the serotonin concentration in the brain; 2) unilateral division of the bundle causes a decrease in the serotonin concentration only in the half of the brain on the side of injury; 3) the time of the fall in the brain serotonin concentration coincides with that of degeneration of the nerve fibers of this tract.

These observations suggest that serotoninergic structures of the hypothalamus belong to the system of the medial forebrain bundle, a conclusion supported by the results of local injection of serotonin into the preoptic region and also by observations made by other workers. Injury to the ascending fibers running along the base of the mesencephalon is accompanied by a marked decrease in the serotonin concentration in the hypothalamus (Parent et al., 1969). These workers conclude that hypothalamic serotonin-containing nerve endings belong to the system of ascending fibers of the medial forebrain bundle.

I have stated several times that experiments in which chemical substances are injected locally into the intact brain cannot rule out the possibility of their indirect action on the pituitary-adrenal system via appropriate peripheral mechanisms. This is shown, in particular, by experiments in which noradrenalin and carbachol were injected locally into the posterior hypothalamus and rostral portions of the mesencephalon of guinea pigs after mesencephalic section (Naumenko, 1967a, 1968). So far as the effect of serotonin is concerned, it was therefore necessary to differentiate between its direct central action and its possible effect on the pituitary-adrenal complex via peripheral mechanisms (Naumenko, 1968; Naumenko et al., 1968).

Injection of Serotonin into the Hypothalamus after Mesencephalic Section. In all experiments the plane of section was far caudally to the cannula. In the series of experiments the sections were of the "cerveau isolé" type, and also pretrigeminal and post-trigeminal, in which case the plane of section passed posteriorly to the inferior colliculi and pons. Just as in the experiments in which serotonin was injected into the lateral ventricle, the type of section had no significant effect either on the initial blood corticosteroid level or on the results, which were similar after section of the brain stem at different levels. The results obtained after the different types of section could accordingly be interpreted as synonymous.

One hour after mesencephalic section serotonin was injected into the medial mammillary nucleus, the ventromedial nucleus of the central hypothalamus, the anterior hypothalamic nucleus, and the preoptic region. Definite stimulation of adrenocortical function was observed 2 h after section (1 h after injection). Meanwhile, injection of serotonin under analogous experimental conditions into

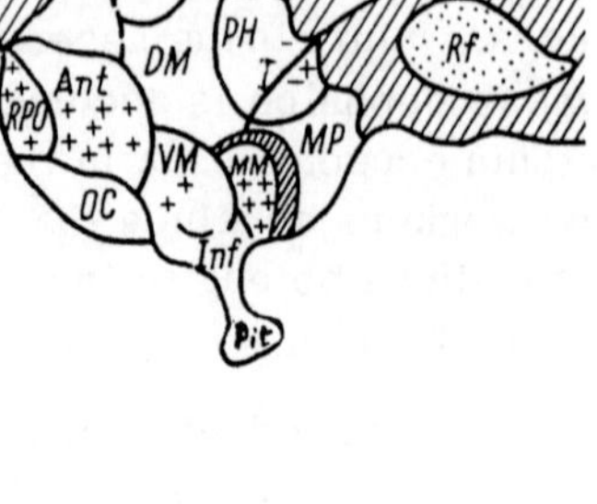

Fig. 20. Scheme of hypothalamus in the sagittal plane. Stimulant (+) and inhibitory (I) effect of serotonin, no response to serotonin (-). RPO) Preoptic region; Ant) anterior hypothalamus; DM) dorsomedial nucleus; VM) ventromedial nucleus; OC) optic chiasma; Inf) infundibulum; Pit) pituitary gland; PH) posterior hypothalamic nucleus; MM) medial mammillary nucleus; MP) posterior mammillary nucleus; Rf) mesencephalic reticular formation; Pv) paraventricular nucleus; Aq) aqueduct of Sylvius.

the posterior hypothalamic nucleus and, in one case, into the supramammillary region did not lead to an increase in the blood corticosteroid level (Table 18, Fig. 20).

Stimulation of the function of the hypothalamo-hypophyseo-adrenal system by serotonin injected locally 1 h after mesencephalic section is evidence that the hypothalamic region contains serotoninergic neurons which can have a direct central action on the hypothalamic zone producing corticotrophin-releasing factor. This is a particularly interesting fact: if the substance in fact influences this system centrally it ought to continue to have a stimulant effect even after transmission of impulses to and from the periphery is made impossible, i.e., in cases in which the substrate on which the substance acts is found in cells belonging to the system of ascending neurons. (Fig. 21).

It will be recalled that in similar experiments neither adrenalin nor carbachol activated the pituitary-adrenal system. Meanwhile, the same effect was observed after injection of these biologically active substances into the posterior hypothalamus of animals with an intact brain.

The question may arise whether this amine activates the pituitary-adrenal complex through changes in the cerebral circulation, for the effects of serotonin on the peripheral circulation are well known (Erspamer, 1966). I consider that such a possibility is unlikely.

It has recently been shown that serotonin, if injected into the

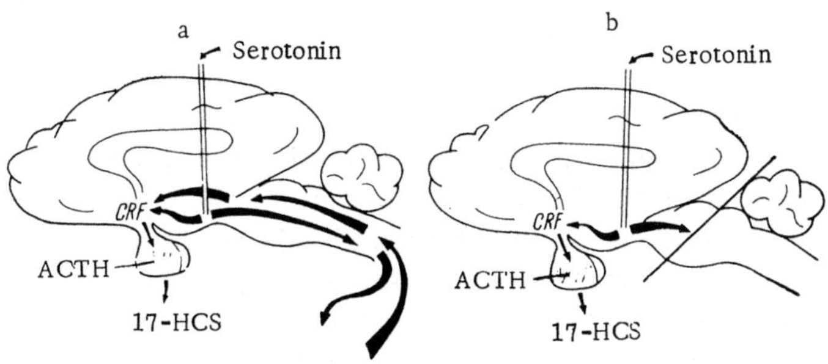

Fig. 21. Scheme showing effect of serotonin, injected locally into the brain, on the hypothalamo-hypophyseo-adrenal system. Serotonin has a direct central action on the hypophyseal zone of the hypothalamus, which produces CRF, for its stimulant effect (a) is preserved after mesencephalic section (b).

hypothalamus or lateral ventricle, increases the blood flow in the hypothalamus only if large doses are given (Hassler and McCook, 1971; Rosendorff and Cranston, 1971). If, however, serotonin was injected in a dose of 5 μg (calculated as the base), no changes in the hypothalamic blood flow were observed (Rosendorff and Cranston, 1971). Meanwhile, serotonin injected locally into the hypothalamus has a stimulant effect on the pituitary-adrenal system in a dose of 4.3 μg (Naumenko, 1968).

Presumably the serotoninergic structures of the hypothalamus which stimulate the pituitary-adrenal system, unlike noradrenalin and acetylcholine receptors, belong at least in part to the system of ascending neurons. I have deliberately used the term "in part," for injection of serotonin into the posterior hypothalamic nucleus and supramammillary region in my experiments after brain section was not accompanied by any increase in adrenocortical function.

The absence of a response of the pituitary-adrenal complex after injection of serotonin into the nuclei mentioned above cannot evidently be explained by inhibition of its functional activity by the brain sections. If, for example, the response of the adrenal cortex to injection of serotonin into the posterior hypothalamic nucleus and into the medial mammillary nucleus of guinea pigs with an intact brain is compared, the degree of elevation of the blood corticosteroid level in the first case is higher than in the second (by 120 and 90% respectively on the average). Meanwhile, injection of serotonin into the posterior hypothalamic nucleus after mesencephalic section did not stimulate adrenocortical function, whereas its injection into the medial mammillary nucleus under the same conditions led to a significant increase of 50% in the corticosteroid concentration in the peripheral blood. The same conclusion can be reached with regard to the supramammillary region compared with the region of the medial mammillary nuclei: the blood corticosteroid level in guinea pigs with an intact brain rose on the average by 160%, whereas after brain section no evidence of stimulation was found. However, after mesencephalic section the blood corticosteroid level rose after injection of serotonin to a lesser degree than in the animals with an intact brain. It could be supposed that this was due to some degree of inhibition of pituitary-adrenal function by the slightly increased initial corticosteroid level (G. Sayers and M. Sayers, 1948) [although this level was far from maximal], as was observed in my experiments with brain section. However, this did not seem a likely cause. There are many facts which justify this statement.

I shall give only two as examples. The blood corticosteroid concentration was increased equally (by 50%) after injection of serotonin into the anterior hypothalamus and the medial mammillary nucleus after brain section, despite the fact that the initial level in the first case was 20% higher. The initial corticosteroid level after injection of noradrenalin (see Chapter II) into the ventral tegmentum and of serotonin into the medial mammillary nucleus was the same (mean values 75.4 and 77.7 μg %). Nevertheless, the corticosteroid level rose on the average by 50% after injection of serotonin under similar conditions, whereas injection of noradrenalin did not change the blood corticosteroid concentration after brain section (or even lowered it by 12%).

The slightly raised blood corticosteroid level, like the operation itself, thus did not play a significant role and could not be the cause of the lesser degree of stimulation of the pituitary-adrenal system after local injection of serotonin in animals with mesencephalic section.

This phenomenon can evidently be explained on the basis of my hypothesis that serotoninergic structures in the hypothalamus belong to systems of both ascending and descending neurons. The latter predominate in the posterior hypothalamus, especially in the posterior hypothalamic nucleus, and also, evidently, in the supramammillary region. In animals with an intact brain, as well as the direct central action of serotonin through afferent neurons on the hypothalamic zone producing CRF, locally injected serotonin can also have an indirect effect transmitted by descending serotoninergic neurons through the spinal cord, and can involve corresponding mechanisms at the periphery in this response. After brain-stem section, however, only the purely central component of serotonin action on this system remains, and the peripheral effects are eliminated. If this point of view is accepted, the absence of a response of the pituitary-adrenal system in animals with mesencephalic section after injection of serotonin into the posterior hypothalamic nucleus and supramammillary region and the decrease in the response to injection of serotonin into the other nuclei of the hypothalamic region investigated can be explained.

My hypothesis undoubtedly requires more strict experimental verification. However, I would emphasize that the possibility of a descending effect of serotonin through the spinal cord has been proved by experiments in which the cord was divided at the level T2: the serotonin concentration below the level of transection fell (Carlsson et al., 1963). In addition, as experiments using a histochemical fluorescence method have shown, serotonin-containing endings belonging to systems of descending neurons exist in the spinal cord (Dahlström and Fuxe, 1965b), and it is claimed (Carlsson et al., 1962; Anden, 1963, cited by Garattini and Valzelli, 1965) that serotonin may play the role of mediator in them.

Effect of 5-Hydroxytryptophan on the Pituitary-Adrenal System of Animals with Chronic Deafferentation of the Mediobasal Hypothalamus

Despite the fact that the method of mesencephalic section which I have used was adequate to explain the role of central serotoninergic structures, as the corresponding control experiments showed, I considered that it would be desirable to obtain similar evidence under conditions ruling out any possibility of the effect of operative trauma. For this purpose, experiments were carried out (Popova et al., 1972a,b) in which the rats were used 4 weeks after complete deafferentation of the mediobasal hypothalamus by the method of Halász (Halász and Pupp, 1965).

By comparison with animals with an intact brain, in animals with complete isolation of the mediobasal hypothalamus the initial blood corticosterone level was slightly raised, but the increase was not statistically significant. Intraperitoneal injection of 5-hydroxytryptophan was accompanied by a marked rise in the serotonin concentration in the deafferented zone of the hypothalamus and by marked activation of the hypothalamo-adrenal system. The corticosterone concentration in the peripheral blood plasma rose considerably, while the response of the pituitary-adrenal system in rats with the isolated hypothalamus to injection of 5-hydroxytryptophan did not differ significantly from its response to injection of the compound into intact rats (Table 19).

Since the region of the mediobasal hypothalamus was completely isolated from nervous influences, the resulting effect could only be due to direct excitation of the serotonin receptors in the hypophyseotrophic zone of the hypothalamus. Corresponding experiments on hypophysectomized animals showed that neither 5-hydroxytryptophan nor the serotonin formed from it have any direct effect on the adrenals in rats (Popova et al., 1972b). The method of complete deafferentation of the mediobasal hypothalamus used in this investigation under chronic experimental conditions ruled out not only the possibility of an effect of serotonin on the hypophyseotrophic region of the hypothalamus transmitted via the periphery, but also the possibility of involvement of higher levels of the brain in this response.

These results thus fully confirm those obtained in acute experiments with mesencephalic section (Naumenko, 1968), and they show that the mediobasal hypothalamus, i.e., that part of the hypothalamus in which CRF is produced, contains serotonin receptors which are evidently directly connected with the mechanism of isolation of this factor. These serotoninergic structures can be presumed to participate in the regulation of the adrenocorticotrophic function of the pituitary.

The Limbic System of the Telencephalon

The presence of serotoninergic structures related to the function of the pituitary-adrenal system has been investigated in the hippocampus, septum, and amygdala (Naumenko, 1969). This choice was deliberate and was determined both by the close anatomical connections which exist between these structures of the limbic system and the hypothalamus and also by the existence of serotoninergic structures in the hypothalamus which influence pituitary-adrenocortical function.

TABLE 19. Effect of 5-Hydroxytryptophan after Complete Isolation of the Hypothalamus on Serotonin Concentration in the Isolated Region and on Blood Corticosterone Level

Substance injected	Plasma corticosterone concentration ±m, μg%	Serotonin concentration in isolated hypothalamus ±m, μg/g
Physiological saline..........	24.7 ± 1.8 (9)*	1.39 ± 0.33 (8)
	$P < 0.001$	$P < 0.05$
5-Hydroxytryptophan, 100 mg/kg......	95.1 ± 8.5 (11)	5.26 ± 1.42 (6)

*Number of animals.

Hippocampus. Two regions of the posterior part of the hippocampus were studied: the ventral hippocampus and the lower part of the dorsal hippocampus (Fig. 22).

The 17-hydroxycorticosteroid concentration in the peripheral blood plasma of the guinea pigs was increased 1 h after injection of serotonin into the ventral hippocampus by more than 90% of the level observed after the injection of physiological saline. The opposite effect was found after injection of serotonin into the dorsal hippocampus: the blood corticosteroid concentration in one guinea pig was unchanged, while in the rest it was lowered by 15 to 75%. In this group of animals as a whole the corticosteroid concentration in the peripheral blood plasma was lowered by 20%, although the difference between their concentration and its initial level was not statistically significant (Fig. 22, Table 20).

These results are contradicted by later experiments in which implantation of serotonin crystals into the ventral or dorsal hippocampus was not followed by changes in pituitary-adrenal function (H. Krieger and D. Krieger, 1970). It is difficult at present to explain this contradiction.

Septum. After injection of serotonin into the septal region, the blood corticosteroid concentration rose. In this series of experiments the mean increase in the peripheral blood hormone con-

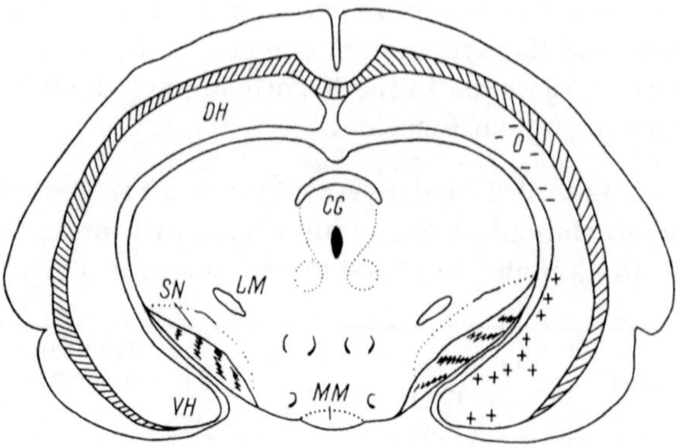

Fig. 22. Diagram of the posterior hippocampus. Stimulant (+) and inhibitory (-) effects of serotonin. No response to serotonin (0). DH) Dorsal hippocampus; VH) ventral hippocampus; MM) medial mammillary nucleus; CG) central gray matter; LM) medial lemniscus; SN) substantia nigra.

centration was 74% (Fig. 23, Table 20). Other workers (H. Krieger and D. Krieger, 1970) have observed an activating effect after injection of serotonin into the septum.

Amygdala. Injection of serotonin into the amygdala evoked different responses of the pituitary-adrenal system in different animals: in most guinea pigs the peripheral blood corticosteroid level fell by more than 20% (from 83 to 66). In three animals the blood corticosteroid concentration was unchanged by comparison with its initial level, and finally, in one guinea pig it was increased by 30%. It is interesting to note that in the experiments of H. Krieger and D. Krieger (1970) an activating effect after injection of serotonin into the amygdala was also observed in only 1 of 11 experiments. In my experiments there was a tendency for the blood corticosteroid level in the animals of this group to fall after injection of serotonin, but the difference was not significant (Fig. 23, Table 20). I was unable to connect the difference observed in these experiments between the response of the pituitary-adrenal system and any particular morphological structure of the amygdala. It seems most likely that neurons responding differently to serotonin are distributed diffusely in the amygdala.

The results of experiments in which serotonin was injected into the anatomical structures of the limbic system show that the hippocampus, septum, and amygdala contain serotoninergic structures whose excitation leads to different effects on pituitary-adrenal activity. The explanation may be that the serotonin receptors belong to different neuronal chains (for example, to ascending and descending chains), or the effects may be brought about by different mechanisms, based on differences in ability to modulate incoming im-

TABLE 20. Blood 17-Hydroxycorticosteroid Concentration after Injection of Serotonin into Structures of the Limbic System in Guinea Pigs

Localization of cannula	Mean plasma corticosteroid concentration ±m, μg/100 ml		P	Number of animals
	Physiological saline	Serotonin		
Hippocampus:				
ventral	62.5 ±5.72	121.8 ±11.69	<0.001	11
dorsal	83.9 ±7.70	66.5 ± 5.08	>0.05	7
Septum	76.0 ±9.64	132.1 ±10.80	<0.01	5
Amygdala	50.9 ±5.42	38.6 ±6.30	>0.1	11

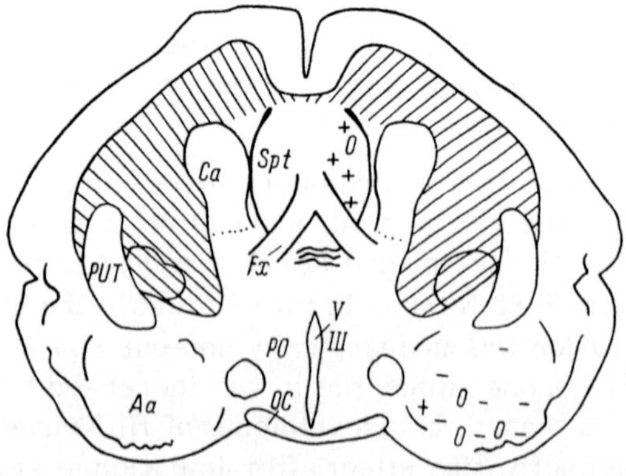

Fig. 23. Diagram of the region of the septum and amygdala. Activating (+) and inhibitory (-) effects of serotonin. No response to serotonin (0). Aa) Amygdala; Spt) septum; Fx) fornix; PO) preoptic region; OC) optic chiasma; V III) third ventricle; PUT) putamen; Ca) caudate nucleus.

pulses. This phenomenon is difficult to explain and requires special investigation. At this stage, however, the fact can be stated that serotoninergic structures connected with the function of the pituitary-adrenal system exist in the limbic formations of the brain.

The experiments in which serotonin was injected into the dorsal hippocampus gave results in harmony with those obtained by many other workers who observed inhibition of the function of the pituitary-adrenal system after electrical stimulation and activation of its function after destruction of the hippocampus (Endröczi et al., 1959; Okinaka et al., 1960a,b; Knigge, 1961; Mandell et al., 1963; Rubin et al., 1966). However, after electrical stimulation of the hippocampus the level of adrenocortical function may actually be raised if the parameters of the electric current are suitably chosen (Endröczi and Lissak, 1962). Admittedly these observations apply to the dorsal part of the hippocampus. Nevertheless, the possibility cannot be ruled out that the ventral hippocampus may be implicated in the activating action on the pituitary-adrenal system observed in these experiments.

The existence of serotoninergic structures in the hippocampus can be postulated, for serotonin has been found in this structure although not in such large amounts as in the hypothalamus (Garattini and Valzelli, 1965), and after microelectrophoretic injection of serotonin there was a decrease both in the spontaneous activity of the hippocampal neurons and in their activity evoked by glutamic acid (Herz and Nacimiento, 1965). These last observations have been confirmed by other experiments (Biscoe and Straughan, 1966), which also showed that after microelectrophoretic injection of serotonin besides the inhibition of spontaneous activity in certain hippocampal pyramidal neurons, the firing rate of others was raised. It can accordingly be concluded that the hippocampus contains serotoninergic structures belonging to qualitatively different neurons. Excitation of some leads to the appearance of inhibitory effects from the hippocampus, while excitation of others leads to an activating effect.

The results of my experiments also show that the serotoninergic structures in the hippocampus may be closely connected with pituitary-adrenal function. Injection of serotonin into the ventral hippocampus led to marked activation of the pituitary-adrenal system. The opposite effect was observed after injection of serotonin into the dorsal hippocampus. It can be concluded from these experiments that the hippocampus contains serotonin-sensitive neurons whose excitation is accompanied by inhibition or activation of pituitary-adrenal function.

A similar pattern is observed in the amygdala, although the results so far obtained suggest rather that the amygdala contains principally serotoninergic structures which inhibit the pituitary-adrenal system. There is little purpose, however, in discussing the existence of serotonin-sensitive neurons in the amygdala stimulating pituitary-adrenal function, on the basis of only the one observation in my experiments.

Nevertheless this possibility cannot be ruled out, for electrical stimulation of the amygdala has been shown to be accompanied by evidence of stimulation of pituitary-adrenal function (Mason, 1958, 1959a; Endröczi et al., 1959; Okinaka, 1961; Rubin et al., 1966). Meanwhile, other workers found no changes in pituitary-adrenal function (Setekleiv et al., 1961) or inhibition (Slusher and Hyde, 1961b) after electrical stimulation of certain zones in the baso-lateral part of the amygdala. I was unable to establish precisely whether inhibitory structures or neurons insensitive to serotonin exist in certain areas. Most probably their distribution is diffuse, although further experiments are necessary before reliable conclusions can be drawn.

Endröczi and Lissak (1960) showed that destruction of the rostral part of the septum is accompanied by stimulation, and of the caudal parts by inhibition of corticosteroid secretion. According to Ahren (1962), destruction of the posteroventral zones of the septum blocks the response of the adrenals to stress. In my experiments serotonin was injected into the caudal portions of the septum and stimulated the pituitary-adrenocortical system. My results therefore agreed with these other observations, and the activation of adrenocortical function observed after electrical stimulation of the septal region (Katsuki, 1961; McHugh et al., 1966) can be regarded, at least in part, as the result of excitation of serotoninergic structures.

The evidence at present available is insufficient to give a clear picture of the connections between serotoninergic neurons of the hippocampus and amygdala and the hypophyseal zone of the hypothalamus. However, the experiments in which serotonin was injected into the septum suggest that neurons containing structures sensitive to serotonin extend from the ventral hippocampus, through the septum, into the preoptic region where they join the medial forebrain bundle. This bundle, in turn, receives fibers from the medial nuclei of the hypothalamus. The work of Heller and Moore (1965) showed that destruction of the septal region is followed by degeneration of descending fibers connected with the medial forebrain bundle and leads to a decrease in the serotonin concentration in the brain.

Since anatomical connections exist through the system of fibers of the fornix between the mammillary bodies and hippocampus (Nauta, 1960), it can also be postulated that the hippocampal serotonin-containing neurons are connected with similar neurons of the medial mammillary bodies which, like the ventral hippocampus, also contain serotoninergic structures linked with the function of the pituitary-adrenal complex (Naumenko, 1968).

Finally, the hippocampal serotonin-containing neurons may be connected through the septum and habenula with the corresponding neurons of the mesencephalon, for injection of serotonin into mesencephalic formations is also accompanied by activation of the adrenal cortex, as will be discussed below.

Mesencephalon

Mesencephalic serotoninergic structures and their relations with pituitary-adrenal activity have been studied in animals by the insertion of cannulas into three different anatomical formations: the reticular formation, the ventral tegmentum, and the habenulo-interpeduncular tract.

Injection of serotonin into the rostral portions of the mesencephalon reticular formation led to marked stimulation of pituitary-adrenal function. The corticosteroid level in the peripheral blood plasma 1 h after injection was increased on the average by 170%.

Similar results were obtained after injection of serotonin into the ventral tegmentum mesencephali, when the blood glucocorticoid level was raised by 124%. Meanwhile, injection of serotonin into the habenulo-interpeduncular tract was not accompanied by any change in adrenocortical function (Fig. 24, Table 21).

The general conclusion can be drawn from the facts described in this section of the present chapter that serotonin, when injected locally into the mesencephalic reticular formation and ventral tegmentum, i.e., into areas with serotonin-containing neurons (Dahlström and Fuxe, 1965a), stimulates the pituitary-adrenocortical system. Consequently, the serotoninergic structures of the mesencephalon presumably must participate in the regulation of pituitary-adrenocortical activity.

The presence of serotonin and its possible role in mesencephalic functions are confirmed by several facts. For example, serotonin has been found in large quantities in the mesencephalon. In the table given by Crawford (1958) the mesencephalon lies in fourth place for serotonin concentration after the hypothalamus, caudate nucleus, and area postrema. In addition, after some me-

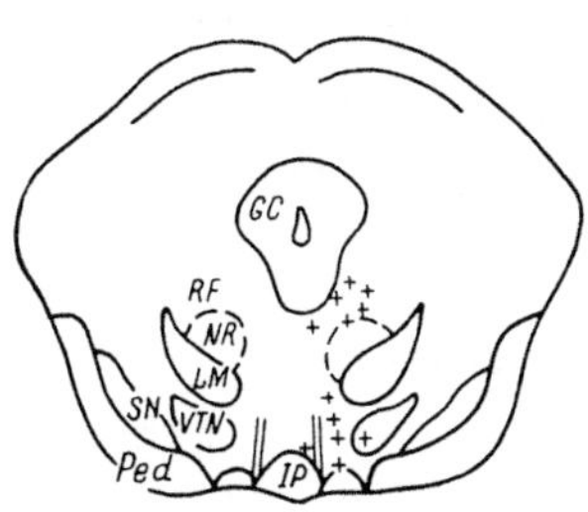

Fig. 24. Diagram of rostral portion of mesencephalon. Activating effect of serotonin (+). GC) Central gray matter; RF) mesencephalic reticular formation; NR) red nucleus; LM) medial lemniscus; VTN) ventral tegmental nucleus; SN) substantia nigra; Ped) cerebral peduncles; IP) interpeduncular nucleus.

TABLE 21. Blood 17-Hydroxycorticosteroid Concentration in Guinea Pigs after Local Injection of Serotonin into the Mesencephalon

Localization of formula*	Mean plasma corticosteroid concentration ±m, μg/100 ml		p	Number of animals
	Physiological saline	Serotonin		
RF	37.1± 6.31	99.5±21.97	<0.05	6
VT	36.8± 1.64	83.3±12.81	<0.02	6
HP	56.2±11.11	58.0±17.94	>0.1	3

*RF) Mesencephalic reticular formation; VT) ventral tegmentum mesencephali; HP) habenulo-interpeduncular tract.

ticulous histochemical investigations Dahlström and Fuxe (1965a) conclude that serotonin-containing neurons are distributed in the region of the central gray matter and reticular formation of the mesencephalon and in the ventral tegmentum. Finally, an investigation conducted by Il'yuchenok (1965) demonstrated the role of mesencephalic serotonin-sensitive structures in the activity of the ascending activating reticular system. My own experiments show that the mesencephalic serotoninergic structures participate in the activity of the pituitary-adrenocortical system.

Since the mesencephalon is closely connected with the hypothalamus, it may be supposed that serotonin-containing neurons of the mesencephalon and hypothalamus are interconnected. On the other hand, the work of Dahlström and Fuxe (1965a) shows that serotonin-containing endings are numerous in the limbic zone of the mesencephalon. It is therefore not impossible that neurons containing serotoninergic structures are connected with similar neurons located in other parts of the limbic system. This hypothesis is suppored by the results of the experiments described above in which serotonin was injected locally into the hippocampus, septum, and amygdala.

It can therefore be concluded that anatomical structures in various parts of the brain – mesencephalon, diencephalon, and telencephalon – contain serotoninergic structures linked with the activity of the pituitary-adrenocortical system. These serotonin-sensitive structures may have activating or inhibitory effects on that system. The fact that the activating effect on the adrenal cortex persists after excitation of the serotoninergic brain structures in acute experiments after mesencephalic section and in chronic experiments after isolation of the medio-basal hypothalamus, when effects transmitted via the periphery are blocked, suggests that these serotoninergic structures are concerned with the direct central regulation of the pituitary-adrenal complex. Results showing that the activating effect persists after total isolation of the hypophyseotrophic zone also confirm my hypothesis (Naumenko, 1968) that one of the terminal neurons stimulating secretion of CRF may be serotoninergic in nature.

Conclusion

A leading role has now been established for the hypothalamus in the regulation of pituitary-adrenal activity. Through the efforts of many investigators there is now solid proof that this part of the brain contains the first humoral link in the mechanism of secretion of adaptive adrenocortical hormones. The important role of the whole ventromedial hypothalamus, from the optic chiasma to the mammillary bodies inclusive, in the mechanism regulating the secretion of the hormone-stimulating pituitary adrenocorticotrophic function has been discovered.

The leading role of the hypothalamic region is due to its strategic position, with its anatomical and functional connections with the telencephalon and mesencephalon. The close relations between the limbic structures of the mesencephalon, the mesencephalic reticular formation, and the hypothalamus on the one hand, and between the hypothalamus and limbic system of the telencephalon and its cortex on the other hand, provide a morphological basis for functional contact between the reticular, limbic, and endocrine systems. Much research has shown that electrical stimulation or extirpation of anatomical structures of the brain is accompanied by definite changes in the functional activity of the pituitary-adrenal system. The mesencephalon, in the region of the tegmentum and reticular formation, contains neurons connected with the function of the hypothalamo-hypophyseo-adrenal complex. Excitation of some of these neurons stimulates, while excitation of others inhibits, its activity.

Structures of this type are contained in the limbic system of the telencephalon. Finally, the neocortex has also been shown to influence pituitary-adrenal function.

The fact that within a comparatively short time so many parts of the brain have been shown to participate in the regulation of the pituitary-adrenal complex is evidence of considerable progress in this field of endocrinology. However, the question of the neurohumoral mechanisms linking these various parts of the brain with the hypophyseal zone of the hypothalamus, where CRF is secreted, requires an answer. In other words, the nature of the hypothetical mediators of the central nervous system which can transmit the nervous impulses carrying information about the secretion of this humoral principle must be identified.

During the last few years the role of acetylcholine, noradrenalin, and serotonin in the transmission of nervous impulses in the brain has been actively discussed. The experimental results described in the appropriate chapters are evidence of considerable progress in this field of neurophysiology. From the facts already obtained, some discussion is possible on the role of adrenergic, cholinergic, and serotoninergic structures in the regulation of hypothalamo-hypophyseo-adrenal function (see Chapters II-IV).

Despite many experiments in which adrenalin or noradrenalin was injected by the usual routes into animals with an intact brain, not to mention observations during parallel studies of the secretion of catecholamines and corticosteroids, it was impossible to determine how these biogenic amines exert their influence on the hypothalamo-hypophyseo-adrenal complex.

Meanwhile, the work of Giuliani et al. (1961) and the results of my first experiments suggested that the essential part of the action of catecholamines on pituitary-adrenal function is their effect not on the central, but on the peripheral adrenergic structures. This hypothesis was put forward after the discovery that naphazoline, a drug with no central effect in the doses used, and amphetamine, affecting both central and peripheral adrenergic structures, led to definite stimulation of the pituitary-adrenal system. Meanwhile pipradol, a sympathomimetic with central action, was almost without effect on adrenocortical function.

These experiments hinted that central adrenergic structures do not participate in activation of the pituitary-adrenal system. However, before this conclusion could be drawn it was necessary to separate the effects of central and peripheral adrenergic structures by a more reliable method.

In my view, before a method of chronic isolation of the mediobasal portion of the hypothalamus was developed, the only method enabling the central and peripheral effects to be separated reliably enough was the method of mesencephalic section, which is widely used in neurophysiological research. However, since mesencephalic section is a somewhat crude procedure, it will be worthwhile briefly mentioning the principal evidence why this method is adequate for the study of the problems requiring solution.

1. After mesencephalic section, although the blood corticosteroid level rose, it reached nothing like its maximal value, nor did it exceed the normal limits pertaining when the mechanisms of stress are studied in guinea pigs. For instance, Yudaev and Afinogenova (1962) give a normal mean blood corticosteroid level of 46-84 μg% for guinea pigs with an intact brain, and in my experiments after mesencephalic section the value was 54-86 μg%.

2. Functional activity of the pituitary-adrenal system was preserved after section, as shown by the experiments described below: a) injection of cold physiological saline into the lateral ventricle was accompanied by stimulation of the adrenal cortex; b) similar results were obtained in animals with brain section after injection of serotonin either into the lateral ventricle or locally into the hypothalamus. The corticosteroid level was increased by 1.5-2 times.

3. The functional activity of the mesencephalic reticular formation and posterior hypothalamus, i.e., of parts of the brain situated closest to the plane of section, also was preserved in the animals with brain section. The evidence for this was as follows: a) EEG activation of the cortex after subcutaneous injection of amphetamine, galanthamine, and neostigmine; extirpation or inhibition of the mesencephalic reticular formation is known to block the activation reaction (Il'yuchenok, 1965); b) preservation of the stimulant effect on the pituitary-adrenal complex after local injection of serotonin into the posterior hypothalamus.

The hypothesis that peripheral adrenergic structures play a leading role was confirmed by experiments which demonstrated the absence of an activating effect of naphazoline and amphetamine, when injected subcutaneously into animals with mesencephalic section, i.e., when the experimental conditions were such that humoral connections remained intact between the periphery and the brain but the nervous connections were interrupted. The section blocked the influence of peripheral adrenergic structures on the hypothalamo-hypophyseo-adrenocortical system, but central effects from the adrenergic brain structures, which remained functionally active, were still possible.

The same pattern was revealed by investigations of the role of central cholinergic structures: an activating effect on adrenocortical function after injection of anticholinesterase drugs was observed only if nervous connections between the periphery and brain remained intact. After mesencephalic section neither galanthamine nor neostigmine had any effect on the pituitary-adrenal system, although they were able to penetrate by the humoral route above the plane of section, where they inhibited acetylcholinesterase activity rostrally to the level of section and induced marked EEG-activation. These experiments thus show that the function of the hypothalamo-hypophyseo-adrenal complex is stimulated only after excitation of peripheral cholinergic structures. Their results agree with those obtained by other workers (Ryzhenkov, 1959a,b; Giuliani et al., 1961; Poskalenko, 1965). At the same time, these results exclude a direct effect of acetylcholine on the adrenal cortex.

Consequently, there was every reason to accept that activation of the function of the pituitary-adrenal system was connected with excitation of peripheral adrenergic or cholinergic structures.

Meanwhile, there was some indirect evidence that a change in the tone of central adrenergic or cholinergic structures could cause stimulation of the hypothalamo-hypophyseo-adrenal complex. I refer to the results showing a change in the noradrenalin and acetylcholine concentrations in the brain after exposure to stress situations (Elliot et al., 1950; McIntosh and Oborin, 1953; Smirnov, 1957; Freedman et al., 1962; Levi and Maynert, 1962; Maynert and Levi, 1964; Éskin and Shchedrina, 1964, 1966, 1968; Gordon et al., 1966; Ordy et al., 1966; Shalyapina, 1967a,b; Kulinskii, 1968; Riege and Morimoto, 1970; Shalyapina and Rakitskaya, 1971; Shchedrina, 1971), indicating some sort of connection between these biologically active substances and the ensuing response of the pituitary-adrenal system. Further evidence of such a connection is given by experiments in which adrenalectomy was followed by lowering of the noradrenalin concentration in the hypothalamus (Shchedrina, 1971), presumably in connection with intensification of its metabolism

(Fuxe et al., 1970a). Injection of cortisol into adrenalectomized rats restores the normal level of noradrenalin metabolism (Fuxe et al., 1970a). Finally, the possibility that adrenergic and cholinergic brain structures may participate in the regulation of pituitary adrenocorticotrophic function is shown by experiments in which stimulation of the adrenal cortex occurred after local injection of carbachol and noradrenalin into the posterior parts of the hypothalamus and rostral part of the mesencephalon in cats (Endröczi et al., 1963; D. Krieger and H. Krieger, 1965; H. Krieger and D. Krieger, 1970).

It thus became necessary to discover why there was this discrepancy between the results I have just described and the established fact that subcutaneous injection of adrenomimetics and anticholinesterase drugs does not stimulate adrenocortical function in animals with mesencephalic section (Naumenko, 1967c,d,e).

This seems to be the right place to emphasize once more that neither the indirect evidence, nor experiments in which noradrenalin and carbachol were injected locally into the brain, causing activation of the pituitary-adrenal complex in cats and in guinea pigs with an intact brain, shed any light on the way in which central adrenergic and cholinergic structures participate in the regulation of the activity of this system.

After excitation of chemically sensitive brain structures, impulses can reach the hypophyseal zone of the hypothalamus, which secrete CRF, in two ways: by the direct pathway along the system of afferent neurons, and indirectly, when elevation of the tone of the central adrenergic or cholinergic structures is able to modify the function of the hypothalamo-hypophyseo-adrenal system secondarily through descending tracts and corresponding peripheral mechanisms.

The indirect mechanism of the activating effects of central adrenergic and cholinergic structures attracted my attention not only in connection with my own results, but also with those obtained by other workers. For instance, Anderson and Holgerson (1966) found a high concentration of noradrenalin in the spinal cord, into which the noradrenalin-containing neurons of the brain stem send many of their endings (Dahlström and Fuxe, 1965b; Fuxe, 1965). Some of these endings innervate mainly the sympathetic trunks (Hillarp et al., 1966b). The cholinergic nature of the transmission of nervous impulses in the spinal cord has also been described (Eccles et al., 1954; Curtis and Eccles, 1958; Curtis, 1961; Way and Sutherland, 1963; Waldron, 1969), and descending pathways of cholinergic nature running from the brain stem into the spinal cord have been demonstrated (Hickey and Barnes, 1971).

It is well known that the hypothalamus is a central regulator and integrator of autonomic functions (Sager, 1962; Tonkikh, 1968). Excitation of its adrenergic structures leads to changes in respiration, muscle tone, vascular tone, and blood pressure and lowers the body temperature (Share and Melville, 1963; Feldberg, 1966; Varagic et al., 1966; Toda et al., 1969; Hassler and McCook, 1971). Electrical stimulation of the posterior hypothalamus causes elevation of the arterial pressure and contraction of the eyelids in rats. These sympathetic responses are weakened by adrenolytics (Morpurgo, 1968). The work of Tonkikh (1968) further showed that acetylcholine can produce reflex stimulation of adrenalin secretion by the adrenal medulla through the hypothalamus; other experiments (Varagic et al., 1966) showed that physostigmine changes the arterial pressure by its effect on central cholinergic structures. It has also been shown that local injection of carbachol into the rostral hypothalamus leads to a decrease in body temperature (Kirkpatrick and Lomax, 1970), while injection of carbachol and of acetylcholine into the posterior hypothalamus is followed by lowering of the systemic arterial pressure (Brezenoff and Jenden, 1969).

There is no doubt that the reactions of sympathetic or parasympathetic nature arising at the periphery and leading to changes in homeostasis could be the basis for stimulation of the hypophyseo-adrenal system, which is highly sensitive to such disturbances (Ramey and Goldstein, 1957). These concentrations made it essential to find a means of differentiating between the ability of central adrenergic and cholinergic structures to exert their activating effects both directly and indirectly.

Appropriate experiments showed that local injection of noradrenalin or carbachol into the posterior hypothalamus and rostral portions of the mesencephalon in animals with mesencephalic section no longer gave rise to signs of stimulation of the hypothalamo-hypophyseo-adrenal complex in contrast with their injection into animals with an intact brain. The distinguishing feature of the action of noradrenalin and carbachol when injected locally into the brain was thus that they activated the function of the pituitary-adrenal system only if the nervous connections between the hypothalamus and mesencephalic reticular formation, on the one hand, and lower levels of the brain and the periphery, on the other hand, were intact. The essential importance of these experimental results is that they proved that the effect of excitation of central adrenergic or cholinergic structures on the hypothalamo-hypophyseo-adrenal system is mediated only through the corresponding peripheral mechanisms.

The results of experiments in which noradrenalin and carbachol were injected locally into animals after mesencephalic section confirmed my conclusion, based on experiments involving subcutaneous injection of amphetamine and anticholinesterase preparations under analogous experimental conditions, regarding the leading role of peripheral adrenergic and cholinergic structures in the regulation of the pituitary-adrenal complex. Division of most of the ascending and descending pathways (excluding the first and second pairs of cranial nerves) prevented amphetamine, galanthamine, and neostigmine from acting on the pituitary-adrenal system not only through reflex excitation of the corresponding peripheral receptors, but also indirectly through peripheral mechanisms after excitation of central adrenergic and cholinergic structures located above the plane of brain section by the drugs injected.

The way in which a change in the concentration of catecholamines and acetylcholine in the central nervous system under the influence of stress situations is reflected in pituitary adrenocorticotrophic function can be pictured on the basis of these results.

A change in the brain noradrenalin or acetylcholine level is evidently accompanied by activation of the corresponding adrenergic or cholinergic structures. However, as the mesencephalic section experiments showed, the central adrenergic and cholinergic structures have no direct effect on CRF. Excitation of these structures is transmitted along descending nerve fibers to the periphery. In turn, activation of the peripheral adrenergic and cholinergic structures leads to the generation of impulses which reach the central nervous system in large numbers and stimulate the hypothalamo-hypophyseo-adrenal complex through biologically active substances other than noradrenalin and acetylcholine.

As has already been mentioned, excitation of central cholinergic structures is accompanied by increased secretion of adrenalin by the adrenal medulla (Tonkikh, 1968). Some workers (Long, 1956; Vogt, 1960; Shreiberg, 1966) consider that this amine, in turn, penetrates into the hypothalamus and stimulates mechanisms which activate pituitary adrenocorticotrophic function.

Meanwhile, it has been shown (Vernikos-Danellis and Marks, 1962; Vernikos-Danellis, 1964) that the blood ACTH concentration is raised as early as 1-2 min after the beginning of stress, including after intravenous injection of adrenalin (Farrell and McCann, 1952). If labeled adrenalin and noradrenalin are injected intravenously, 2 min later they have still not penetrated into the hypothalamic neurons (Weil-Malherbe et al., 1959; Weil-Malherbe, 1960). These findings are evidence against a role of adrenalin of the adrenal medulla as a hypothalamic triggering factor, activating the function of the pituitary-adrenal system.

This, of course, does not mean that adrenalin has no role to play in the development of the general adaptation syndrome. There is evidence that in certain types of stress (ether anesthesia, x-ray irradiation) adrenalin is discharged into the blood stream fairly rapidly (Kulinski, 1961, 1965, 1968; Utevskii and Kulinski, 1967). However, the advocates of the humoral effect of adrenalin on hypothalamic mechanisms of regulation of ACTH secretion evidently disregard the fact that the increased secretion of this amine by the adrenal medulla, which is observed, moreover, during exposure to by no means all stress situations (Franksson et al., 1954; Hume, 1958; Goldfien and Ganong, 1962; Euler, 1964), should induce in the first place the many and varied effects connected with the increase in tone of the peripheral adrenergic structures produced by adrenalin. The peripheral action of adrenalin is well known and requires no special description. The numerous peripheral effects produced by adrenalin, in my opinion, could be the cause of reflex or neurohumoral stimulation of function of the hypothalamo-hypophyseo-adrenal system.

Naturally I do not by any means imply that no adrenalin whatever penetrates through the blood–brain barrier or that it has no effect on central adrenergic structures. The possibility of its penetration by the humoral route into the brain has been demonstrated by a highly sensitive radioisotope method (Weil-Malherbe et al., 1959). However, it must be remembered that the rate of penetration is such that it rules out any possibility that adrenalin acts as a triggering factor, and my experiments are evidence against a direct central participation of adrenergic structures in the regulation of activity of the hypothalamo-hypophyseo-adrenal complex. These considerations do not mean that the adrenalin of the adrenal medulla is in general incapable of participating in the central regulation of endocrine functions. For example, it has been shown that adrenalin activates hypothalamic mechanisms concerned with the regulation of hormones of the posterior lobe of the pituitary (Tonkikh, 1968).

It is therefore possible that in certain stress situations the adrenalin liberated from the adrenal medulla penetrates into the brain where it excites central adrenergic structures. However, in such cases its action on the pituitary-adrenocortical system is transmitted along descending adrenergic pathways via peripheral mechanisms. The role of central adrenergic and cholinergic structures in the mechanisms of regulation of the adrenocorticotrophic function of the pituitary is thus secondary and is connected with their influence on the corresponding peripheral mechanisms. The study of these mechanisms does not fall within the scope of the present investigations. However, it can be considered that many of the widely different disturbances of homeostatic equilibrium which develop at the periphery as a result of changes in the tone of the autonomic nervous system of central origin (Magoun et al., 1937; Sager, 1960; Share and Melville, 1963; Bernardis and Skelton, 1965; Gann and Egdahl, 1965; Feldberg, 1966; Folkow and Rubinstein, 1966; Schallek and Zabranski, 1966; Varagic et al., 1966; Sheard and Aghajanian, 1967; Morpurgo, 1968; Tonkikh, 1968) are causes of reflex and, evidently, humoral reciprocal effects on the brain. These effects, in turn, stimulate the function of the hypothalamo-hypophyseo-adrenal system through an increase in the tone of central chemically reactive structures which are neither adrenergic nor cholinergic.

When the possible participation of catecholamines in the activity of the pituitary-adrenal system is discussed, it is principally noradrenalin which is considered and not dopamine. The role of dopamine in regulation of the pituitary-adrenocortical system is still in question (Fuxe et al., 1970a), although such a role has been demonstrated for the pituitary–gonads system (Fuxe et al., 1970b; Schneider and McCann, 1970; Kamberi et al., 1971). Whatever the case, injection of dopamine into the lateral ventricle in rats causes no change in the corticosterone concentration in the adrenals (Marks et al., 1970).

Ganong and co-workers have recently published a series of papers (Lorenzen and Ganong, 1967; Van Loon et al., 1967b) which indicate that noradrenalin of the brain under conditions of maximal stress can inhibit the pituitary-adrenal complex. The physiological significance of these results is not yet clear. If noradrenalin did participate in the direct regulation of ACTH secretion, in fact by inhibiting it, the complete disappearance of noradrenalin produced by complete deafferentation of the mediobasal hypothalamus (Weiner et al., 1972) ought to be accompanied by a maximal increase in the blood corticosteroid level. However, after total isolation of the hypothalamus the initial concentration of adrenocortical hormones in the blood is either moderately raised (Halász et al., 1967b; Halász, 1969) or it does not differ significantly from the blood glucocorticoid concentration in animals with an intact brain (Makara et al., 1970; Popova et al., 1972a,b). On the other hand, the response to humoral stressors (Makara et al., 1970) or to 5-hydroxytryptophan (Popova, et al., 1972a,b) after exclusion of the possible inhibitory effect of noradrenalin by total isolation of the hypophyseotrophic zone (Weiner et al., 1972), is not increased by comparison with animals with an intact brain.

Other workers have recently concluded that brain neurons containing catecholamines do not play any essential role in the regulation of the pituitary-adrenal system. Experiments have shown that exhaustion of the brain catecholamine depots by injection of reserpine into the median eminence of the hypothalamus (Smelik, 1967) or by parenteral injection of reserpine followed by α-methyltyrosine, which blocks noradrenalin synthesis (Carr and Moor, 1968; Hirsch and Moor, 1968), did not change the response of the pituitary-adrenal system to stress.

Results published in the literature and my own experiments in which monoamine oxidase inhibitors, reserpine, and p-chlorophenylalanine, which blocks serotonin synthesis, were given provided the basis for the hypothesis that central serotoninergic structures participate in the mechanisms regulating the pituitary-adrenocortical system.

Since 5-hydroxytryptophan, unlike serotonin, passes easily through the blood–brain barrier (Udenfriend et al., 1957a,b; Bogdanski et al., 1958a,b; Costa, 1960), I considered that the stimulation of the adrenal cortex taking place after subcutaneous injection of the serotonin precursor may be due, at least partially, to its conversion into serotonin in the central nervous system. This hypothesis is confirmed by experiments in which the degree of accumulation of serotonin in the brain was found to be directly dependent upon elevation of the blood hydrocortisone level in animals after subcutaneous injection of 5-hydroxytryptophan. Finally, stimulation of the function of the hypothalamo-hypophyseo-adrenal system after injection of serotonin into the lateral ventricle strengthened my conviction that the central serotoninergic structures are linked with the functional activity of that system.

The possibility that serotonin may have a primary central action was shown in experiments in which this amine was injected into the cerebral ventricles of animals after mesencephalic section. These experiments showed that even in the absence of descending nervous influences excitation of the central serotoninergic structures, by contrast with excitation of central adrenergic and cholinergic structures, is accompanied by stimulation of the function of the hypothalamo-hypophyseo-adrenal complex. These experiments demonstrate the direct central effect of serotoninergic structures on the adrenocorticotrophic function of the pituitary, in contrast with the pathways of the activating effect arising after excitation of brain structures sensitive to noradrenalin and acetylcholine (Naumenko, 1965, 1966, 1967a,b,c,d,e, 1968).

The conclusion that the serotoninergic receptors of the brain play a direct part in the regulation of the pituitary-adrenal complex was confirmed by chronic experiments on rats with complete deafferentation of the hypophyseotrophic zone of the hypothalamus (Popova et al., 1972a,b). The fact that when reflex influences from other parts of the brain were completely blocked, intraperitoneal injection of 5-hydroxytryptophan produced an increase in the serotonin level in the isolated insula and marked activation of the pituitary-adrenal complex is firm evidence, in my opinion, that the serotoninergic neurons of the hypophyseotrophic zone of the hypothalamus participate in the regulation of CRF secretion.

Once the direct central effect of serotoninergic structures on pituitary-adrenal activity had been established, the way was open for detection of the actual localization of these structures in the central nervous system. Attention was directed initially to the hypothalamic region, for it is there that nervous impulses arising in response to a wide variety of stressors are switched to humoral influences, namely the secretion of CRF. Experiments showed that the hypothalamic region contains structures sensitive to serotonin, whose excitation is followed by increased pituitary-adrenal activity. They are located throughout the medial part of the hypothalamus from the optic chiasma to and including the mammillary bodies. The lateral zones of the hypothalamic region, according to the results of these experiments, do not contain serotoninergic structures connected with the function of the pituitary-adrenal system.

The distribution of serotoninergic structures in the anterior and posterior hypothalamus is evidently identical, so far as can be judged from the approximately identical increase in the blood corticosteroid level after local injection of serotonin. There are grounds for this hypothesis, for according to Garattini and Valzelli (1965) the serotonin concentration in these zones is identical.

The serotoninergic structures thus identified are evidently connected with different neuronal pathways. One of them belongs to a system of ascending neurons, afferent relative to the hypophyseotrophic zone producing CRF, while the rest belong to a system of descending pathways. This conclusion is drawn from experiments in which serotonin was injected into various zones of the hypothalamus in guinea pigs with mesencephalic section. Under these conditions the serotonin continued to activate the pituitary-adrenal system, but its activating effect was weaker, and after injection of serotonin into areas of the posterior hypothalamus its effect did not occur in all cases, as it did when injected into animals with an intact brain. For example, injection of serotonin after brain section into the posterior hypothalamic nucleus no longer stimulated the function of the pituitary-adrenal system, whereas its injection into the medial mammillary nucleus continues to have an activating effect under the analogous conditions in all experiments.

It is interesting to note that, according to some observations, serotonin is present in large quantities in the spinal cord (Anderson and Holgerson, 1966), and is contained in the endings of descending nerve fibers (Dahlström and Fuxe, 1965b) where, it is claimed, it may play the role of mediator (Carlsson et al., 1962; Anden et al., 1963, cited by Garattini and Valzelli, 1965). Some of the serotoninergic structures in certain posterior hypothalamic nuclei evidently belong almost entirely to a system of descending neurons. All these observations suggest that, besides the direct central effect on the hypothalamo-hypophyseo-adrenal system the serotoninergic structures of the brain may also, like the adrenergic and cholinergic structures, exert an indirect effect through the participation of peripheral mechanisms.

The close anatomical and functional connections of the hypothalamus with the mesencephalon and the limbic system of the telencephalon explain the great interest which is shown in the role of these parts of the brain in the mechanisms regulating the functions of the hypothalamo-hypophyseo-adrenal system. It was not known whether the serotoninergic structures of the telencephalon and mesencephalon play a role in the activity of the pituitary-adrenal complex. However, the search for structures of this type was stimulated by the following facts:

1. The limbic system (hippocampus, amygdala, septum) was shown to contain large quantities of serotonin (Matussek, 1966). Even more serotonin was found in the mesencephalon (Crawford, 1958).

2. Neurons sensitive to serotonin were found in the mesencephalon (Dahlström and Fuxe, 1965a) and in the hippocampus (Herz and Nacimiento, 1965). The main concentrations of serotonin neurons were in fact located in the mesencephalon (Dahlström and Fuxe, 1965a), and it is the neurons of the mesencephalon which are the source of origin for the serotoninergic pathways running to the telencephalon and giving off numerous synaptic branches to the hypothalamus. Experiments have shown that labeled serotonin, injected into the lateral ventricles, accumulates in particular in the nerve endings of the central gray matter of the mesencephalon (Aghajanian and Bloom, 1967).

3. The serotoninergic structures of the mesencephalic

reticular formation have been shown to participate in the activity of the ascending system (Il'yuchenok, 1967).

On the basis of these results there is evidently ample justification for investigating whether the mesencephalon and the limbic structures of the telencephalon contain chemical structures responding to serotonin which might exert their influence on the pituitary-adrenocortical system.

Experiments showed that the ventral tegmentum and mesencephalic reticular formation contain serotoninergic structures whose excitation stimulates pituitary-adrenal activity. These results show that the serotoninergic structures of the mesencephalon are connected with activity of the pituitary-adrenal complex. Similar structures have also been found in the limbic formations of the telencephalon (Naumenko, 1969; H. Krieger and D. Krieger, 1970).

The hippocampus is known to contain serotoninergic structures, for electrophoretic injection of serotonin into it is followed by a decrease in hippocampal unit activity, both spontaneous and induced by glutamic acid (Herz and Nacimiento, 1965). The structures in the hippocampus which react to serotonin evidently belong to qualitatively different groups of neurons, for microelectrophoretic injection of serotonin into the pyramidal cells of the hippocampus inhibits spontaneous activity but at the same time it increases the firing rate of some neurons (Biscoe and Straughan, 1966). This possibility is confirmed by experiments in which injection of serotonin through a microcannula into the ventral hippocampus caused stimulation, while into the inferior portion of the dorsal hippocampus it caused inhibition of the function of the pituitary-adrenal system.

A similar pattern was observed in the amygdala, although the results of my experiments are more in harmony with the view that it contains predominantly serotoninergic structures which inhibit the adrenal cortex. Finally, neurons stimulating this function and sensitive to serotonin have been found in the septal region.

Certain anatomical structures of the limbic system thus contain chemical receptors sensitive to serotonin; excitation of some is accompanied by stimulation while excitation of others is accompanied by inhibition of the function of the pituitary-adrenal system.

These facts are in accordance with the view of Salmoiraghi (1966) that the same mediator can have both an excitatory and an inhibitory action. The direction of the reaction depends on certain characteristics of the neuron concerned. Salmoiraghi himself does not rule out the possibility that there are two types of receptor for the same mediator. Its action on receptors of one type leads to changes accompanied by hyperpolarization of the cell membranes and inhibition of unit activity, while its action on receptors of the other type has the opposite effect – depolarization of the cell membrane and excitation of unit activity. The existence of different types of "tryptamine" receptors (M-receptors, blocked by morphine, and D-receptors, blocked by dibenzyline) was established, in principle, by Gaddum (1961; cited by Garattini and Valzelli, 1965).

It can be postulated from the evidence cited that either the serotonin receptors found in the limbic system belong to different neuronal groups or the direction of their response is determined by the influence of other parts of the central nervous system in contact with particular zones of the hippocampus.

Impulses arising after excitation of the serotoninergic structures of the hippocampus and amygdala evidently reach the hypothalamus via the septum and preoptic region, where the principal pathways connecting these structures of the limbic system with the hypothalamus run (Nauta, 1956, 1958, 1960; Gloor, 1965; Lundberg, 1960). Nevertheless, the problem of how the inhibitory influences are transmitted from the dorsal hippocampus and amygdala to the function of the hypothalamo-hypophyseo-adrenal system remains unexplained. One difficulty is that neither in the septum nor in the preoptic region, the main relay points for nervous impulses passing from the hippocampus and amygdala to the hypothalamus, has it yet been possible to identify serotoninergic structures whose excitation would have an inhibitory effect. I also have not succeeded in finding serotoninergic structures inhibiting the function of the pituitary-adrenal system. Of course this does not mean that there are no serotoninergic structures in the hypothalamic region of this type, and indeed, as Salmoiraghi (1966) considers, two types of receptors sensitive to the same substance may be found on the same nerve cell. However, the data on this problem in the literature (Shreiberg and Dunaeva, 1965; Shreiberg, 1966; Shreiberg and Matlina, 1967), as has already been stated in Chapter IV, are unconvincing.

Another possible explanation seems more likely, namely that nervous impulses are relayed in the dorsal hippocampus and amygdala to nonserotoninergic structures capable of inhibiting the pituitary-adrenocortical system. This role may evidently be played by chemical receptors sensitive to acetylcholine.

Neurons whose spontaneous activity is persistently increased after microelectrophoretic injection of acetylcholine exist in the limbic system, especially in the hippocampus and septum (Herz and Nacimiento, 1965). On the other hand, electrical stimulation of the hippocampus and, in some experiments, of the septum and amygdala also induces inhibition of the function of the hypothalamo-hypophyseo-adrenal system (Endröczi et al., 1959; Slusher and Hyde, 1961b; Endröczi and Lissak, 1963; Lissak and Endröczi, 1965, Rubin et al., 1966).

The same result was obtained 1 h after local injection of carbachol into the region of the septum and preoptic zone (Endröczi et al., 1963). Conjecturally, therefore, the dorsal hippocampus contains serotoninergic structures whose excitation, relayed to structures sensitive to acetylcholine, is transmitted via the septum and preoptic region to the hypothalamus and to its hypophyseal zone which produces CRF, where it causes inhibition of the secretion (synthesis ?) of this factor, and thus inhibits the pituitary-adrenal system. The inhibition of this system observed in my experiments after injection of serotonin into the amygdala can be explained in the same way.

Is serotonin the only substance capable of participating in the direct central regulation of the hypothalamo-hypophyseo-adrenal system ? The facts at present available suggest that this is unlikely. It has been shown, for instance, that after local exhaustion of the monoamine depots in the median eminence of the rat hypothalamus by reserpine the animals continued to respond to stress by an increase in the function of the pituitary-adrenal system (Smelik, 1967). In the experiments of Dixit and Buckley (1969), injection of 4-chloramphetamine into rats or keeping the animals on a diet low in tryptophan led to a decrease in the serotonin concentration in the brain by 74 and 63% respectively. However, against this background, stress (cold and shaking the animals) evoked the same response of the adrenal cortex as it did in control animals. Admittedly these workers suggest that under these conditions the rate of serotonin metabolism may be increased, for stress has been shown to intensify serotonin synthesis (Thierry et al., 1968). In my experiments injection of *p*-chlorophenylalanine, which sharply reduces the serotonin concentration in the brain, weakened the response of the pituitary-adrenal system. Nevertheless, elevation of the blood corticosteroid level after injection of formalin or naphazoline was quite well marked.

Perhaps besides serotonin there are other, as yet unknown substances which participate in the regulation of the hypothalamo-hypophyseo-adrenal system. The possibility cannot be ruled out, for example, that other biologically active substances formed in the brain may participate in the regulation of the function of this system, substances such as γ-aminobutyric acid, dopamine, and prostaglandins, the role of which in this connection is still unknown. It is suggested that these factors may play the role of mediators in the central nervous system (Elliot and Jasper, 1959; Roberts and Eidelberg, 1960; Rossum, 1963; Bertler and Rosengren, 1966; King and Thomas, 1968; Horton, 1969).

Although the facts described above seem to imply that one of the terminal neurons stimulating the secretion and, perhaps, the synthesis of CRF is serotoninergic in nature, the possibility therefore cannot be ruled out that terminal neurons with a different type of chemical sensitivity may exist. Moreover, there are still insufficient grounds for asserting that the action of serotonin is connected with its property of mediator, for its role in the central nervous system as a modulator of nervous impulses must be taken into account (Florey, 1967).

It is perfectly clear that much hard work will be required before the whole picture of the central neurohumoral regulation of the hypothalamo-hypophyseo-adrenal complex, the vital importance of which is sufficiently well known, can be revealed.

Bibliography

Aghajanian, G. K., and Bloom, F.E. (1966). "Electron-microscopic autoradiography of rat hypothalamus after intraventricular H^3 norepinephrine," Science, 153:303-310.

Aghajanian, G. K., and Bloom, F. E. (1967). "Localization of tritiated serotonin in rat brain by electron-microscopic autoradiography," J. Pharmacol. Exptl. Therap., 156:23-30.

Ahlquist, R. P. (1948). "A study of the adrenotropic receptors," Amer. J. Physiol., 153:586-600.

Ahren, C. (1962). "Effect of diencephalic lesions on acute and chronic stress responses in male rabbits," Acta Endocrinol., 39. Suppl. 69:1-92.

Airaksinen, M. M., Giacalone, E., and Valzelli, L. (1968). "Hydroxylation of tryptophan in the brain *in vivo*," J. Neurochem., 15:55-56.

Aleshin, B. V. (1960). "The role of the hypothalamus in the regulation of the anterior pituitary," Uspekhi Sovr. Biol., 49(1):115-135.

Aleshin, B. V. (1964). "The role of neurosecretion in the hypothalamic regulation of endocrine functions," in: Neurosecretory Elements and Their Role in the Organism [in Russian], Nauka, Moscow, pp. 32-72.

Aleshin, B. V. (1971). The Histophysiology of the Hypothalamo-Hypophyseal System [in Russian], Meditsina, Moscow.

Alikmets, L. (1964). "The functional role and pharmacology of the limbic system," Zh. Nevropat. i Psikhiat., 64(8):1241-1248.

Al'pern, D. E. (1963). Cholinergic Processes in Pathology [in Russian], Medgiz, Moscow.

Anand, B. K., and Dua, S. (1955). "Hypothalamic involvement in the pituitary adrenocortical response," J. Physiol. (London), 127:153-156.

Anden, N.-E., Carlsson, A., and Haggendal, J. (1969). "Adrenergic mechanisms," Ann. Rev. Pharamacol., 9:119-134.

Anden, N.-E., Dahlström, A., Fuxe, K., and Larsson, K. (1965). "Mapping out of catecholamine and 5-hydroxytryptamine neurons innervating the telencephalon and diencephalon," Life Sci., 4:1275-1279.

Anderson, R. N., and Egdahl, R. H. (1964). "Chemical mediation of ACTH secretion in the dog," in: Major Problems in Neuroendocrinology, E. Bajusz and G. S. Jasmin (editors), Karger, Basel, pp. 149-155.

Anderson, E. C., and Holgerson, L. O. (1966). "The distribution of 5-hydroxytryptamine and norepinephrine in cat spinal cord," J. Neurochem., 13:479-485.

Anichkov, S. V., and Belen'kii, M. L. (1962). The Pharmacology of the Chemoreceptors of the Carotid Body [in Russian], Medgiz, Leningrad.

Arezio, G., and Peruzzo, L. (1953). "Histochimie de la surrénale du rat sous hexamethonium dans le stress," Anesth. Analg., 10:371-377.

August, S., and Gubner, R. (1949). " Effect of an adrenolytic compound, dihydroergocornine, on epinephrine-induced stimulation of the adrenal cortex," Bull. New York Acad. Med., 25:446-447.

Axelrod, J. (1957). "O-methylation of epinephrine and other catechols *in vitro* and *in vivo,* " Science, 126:400-401.

Axelrod, J. (1959). "The metabolism of catecholamines *in vivo* and *in vitro* ," Pharmacol. Rev., 11:402-408.

Axelrod, J., Albers, W., and Clemente, C. D. (1959). "Distribution of catechol-O-methyltransferase in the nervous system and other tissues," J. Neurochem., 2:68-72.

Axelrod, J., and Tomchik, R. (1958). Enzymic O-methylation of epinephrine and other catechols," J. Biol. Chem., 233:702-705.

Azmitia, E. C., Jr., and McEwen, B. S. (1969). "Corticosterone regulation of tryptophan hydroxylase in midbrain of the rat," Science, 166:1274-1276.

Bacchus, H. (1954). "Influence of epinephrine on the plasma content of 17-hydroxycorticosteroids," Amer. J. Physiol., 178:437-440.

Barnett, S. A. (1963). A Study in Behaviour. Principles of Ethology and Behavioural Physiology, Displayed Mainly in the Rat," Methuen, London.

Benetato, G., Oluitu, M., Bubuianu, E., and Bonciocat, C, (1967). "The effect of sectioning of the medial forebrain bundle and of sodium diethylbarbiturate on the content in catecholamines and serotonin of the hypothalamus and rhinencephalic structures (amygdala and hippocampus)," Rev. Roum. Physiol., 4:13-25.

Berkun, M. M. (1964). "Performance decrement under physiological stress," Human Factors, 6:21-30.

Bernardis, L. L., and Skelton, F. R. (1965). "Blood pressure in female and male rats following ventromedial hypothalamic lesions placed at four different areas," Proc. Soc. Exptl. Biol. (New York), 120:756-760.

Bertelli, A., Cantone, G., and Martini, L. (1954). Azione della serotonin sull'asse ipofisi-surrene," Atti Soc., Lombarda Sci. Med.-Biol., 9:10-12.

Bertler, A., and Rosengren, E., (1959). "On the distribution in brain of monoamines and of enzymes responsible for their formation," Experientia, 15:382-388.

Berthler, A., and Rosengren, E. (1966). "Possible role of brain dopamine," Pharmacol. Rev., 18:769-773.

Besson, M. J., Cheramy, A., and Glowinski, J. (1969). "Effects of amphetamine and desmethylimipramine on amine synthesis and release in central catecholamine-containing neurons," Europ. J. Pharmacol., 7:111-114.

Bhagat, B. (1969). "Effect of chronic cold stress on catecholamine levels in rat brain," Psychopharmacologia (Berlin), 16:1-4.

Bhattacharya, A. N., and Marks, B. H. (1969). "Effect of pargiline and amphetamine upon

acute stress responses in rats," Proc. Soc. Exptl. Biol. (New York), 130:1194-1198.
Biscoe, T. J., and Straughan, D. W. (1966). "Micro-electrophoretic studies of neurons in the cat hippocampus," J. Physiol. (London), 183:341-359.
Bisso, A. (1953). "Comparison of the effects of epinephrine and norepinephrine on circulating eosinophils," J. Clin. Endocrinol. Metabol., 13:1226-1231.
Björklund, A., Falck, B., and Stenevi, U. (1971). "Classification of monoamine neurons in the rat mesencephalon: distribution of a new monoamine neurone system," Brain Res., 32:1-17.
Blair-West, I. R., Cochlan, I. P., Denton, D. A., Goding, I. R., Munro, I. A., Peterson, R. E., and Wintour, M. (1962). 'Humoral stimulation of adrenal cortical secretion," J. Clin. Investig., 41:1606-1627.
Bliss, E. L., Ailion, J., and Zwanziger, J. (1968). "Metabolism of norepinephrine, serotonin and dopamine in rat brain with stress," J. Pharmacol. Exptl. Therap., 164:122-134.
Bliss, E. L., Rubin, S., and Gibert, Th. (1951). "The effect of adrenalin on adrenal cortical function," J. Clin. Endocrinol. Metabol., 11:46-60.
Bloom, F. E., Oliver, A. P., and Salmoiraghi, G. C. (1963). "The responsiveness of individual hypothalamic neurons to microelectrophoretically administered endogenous amines," Internat. J. Neuropharmacol., 2:181-193.
Bloom, F. E., and Giarman, N. J. (1968). "Physiologic and pharmacologic considerations of biogenic amines in the nervous system," Ann. Rev. Pharmacol., 8:229-258.
Bogdanski, D. F., and Udenfriend, S. (1956). "Serotonin and monoamine oxidase in brain," J. Pharmacol. Exptl. Therap., 116:7-8.
Bogdanski, D. F., Weissbach, H., and Udenfriend, S. (1958a). "5-Hydroxytryptophan. Pharmacological studies," J. Pharmacy Pharmacol., 10:525.
Bogdanski, D. F., Weissbach, H., and Udenfriend, S. (1958b). "Pharmacological studies with the serotonin precursor, 5-hydroxytryptophan," J. Pharmacol. Exptl. Therap., 122:182-194.
Bohus, B. (1961). "The effect of central nervous lesions on pituitary-adrenocortical function in the rat," Acta Physiol. Acad. Sci. Hung., 20:373-377.
Bonvallet, M., Dell, P., and Hiebel, G. (1953). "Simus carotiden et activité électrique cérébrale," Compt. Rend. Soc. Biol., 147:1166-1169.
Bouman, P. R., Gaarenstroom, I. H., Smelik, P. G., and de Wied, D. (1957). "Hypothalamic lesions and ACTH secretion in rats," Acta Physiol. Pharmacol. Neerl., 6:368-378.
Braunsteiner, H., Pakesch, F., and Vetter, H. (1951). "Der Eosinophilensturz nach Adrenalin," Klin. Wschr., 63:359-362.
Breitner, C., Picchioni, A., Chin, L., and Burton, L. E., (1961). "Effect of electrostimulation on brain 5-hydroxytryptamine concentration," in: Diseases of Nervous System, Monogr. Suppl., 22:1-4.
Breitner, C., Picchioni, A., and Chin, L. (1964). "Neurohormone levels in brain after CNS stimulation including electrotherapy," J. Neuropsychiat., 5:153-158.
Bremer, F. (1935). "Cerveau isolé et physiologie du sommeil," Compt. Rend. Soc. Biol., 118:1235-1241.

Brezenoff, H. E., and Jenden, D. J. (1969). "Modification of arterial blood pressure in rats following microinjection of drugs into the posterior hypothalamus," Int. J. Neuropharmacol., 8:593-600.

Brodie, B. B. (1958). "Storage and release of 5-hydroxytryptamine – possible significance in chemical mediation in brain," in: 5-Hydrodytryptamine, Lewis (editor), Pergamon Press, Oxford, pp. 64-83.

Brodie, B., and Beaven, M. A. (1963). "Neurochemical transducer systems," Med. Exptl., 8:320-351.

Brodie, B. B., Comer, H. S., Costa, E., and Dlabac, A. (1966). "The role of brain serotonin in the mechanism of the central action of reserpine," J. Pharmacol. Exptl. Therap., 152:340-349.

Brodie, B. B., Spector, S., and Shore, P. A. (1959). "Interaction of monoamine oxidase inhibitors with physiological and biochemical mechanisms in brain," Ann. New York Acad. Sci., 80:609-614.

Brodish, A. (1963). "Diffuse hypothalamic system for the regulation of ACTH secretion," Endocrinology, 73:727-735.

Brodish, A. (1964a). "Delayed secretion of ACTH in rats with hypothalamic lesions," Endocrinology, 74:28-34.

Brodish, A. (1964b). "Role of the hypothalamus in the regulation of ACTH release," in: Major Problems in Neuroendocrinology, Karger, E. Bajusz and G. Jasmin (editors), Karger, Basel, pp. 177-195.

Brodish, A., and Long, C. N. H. (1962). "ACTH-releasing hypothalamic neurohumor in peripheral blood," Endocrinology, 71:298-306.

Bronson, F. H., and Eleftheriou, G. (1965). "Relative effects of fighting on bound and unbound corticosterone in mice," Proc. Soc. Exptl. Biol. (New York), 18:146-149.

Bronson, F. H., and Eleftheriou, B. E. (1965a). "Behavioral, pituitary and adrenal correlates of controlled fighting in mice," Physiol. Zöol., 38:406-411.

Bronson, F. H., and Eleftheriou, B. E. (1965b). "Adrenal responses of fighting in mice: separation of physical and psychological causes," Science, 147:627-628.

Brown, B. B., and Werner, H. W. (1954). "Pharmacologic studies on a new central stimulant, α-(2-piperidyl) benzhydrol hydrochloride (MRD-108)," J. Pharmacol. Exptl. Therap., 110:180-187.

Brown, G. M., Schalch, D. S., and Reichlin, S. (1971a). "Patterns of growth hormone and cortisol responses to psychological stress in the squirrel monkey," Endocrinology, 88:956-963.

Brown, G. M., Schalach, D. S., and Reichlin, S. (1971b). "Hypothalamic mediation of growth hormone and adrenal response in the squirrel monkey," Endocrinology, 89:694-703.

Burger, A. (editor) (1967). Drugs Affecting the Peripheral Nervous System, I., Marcel Dekker, New York.

Burgus, R., and Guillemin, R. (1970). "Hypothalamic releasing factors," Ann. Rev. Biochem., 39:499-526.

Burnstein, Sh. (1952). "The guinea pig as a laboratory animal for corticoid excretion studies: modification of the formaldehydogenic method," Endocrinology, 50: 412-418.

Carlsson, A. (1959). "The occurrence, distribution and physiological role of catecholamines in the nervous system," Pharmacol. Rev., 11:490-493.

Carlsson, A., Falck, B., and Hillarp, N. (1962). "Cellular localisation of brain monoamines," Acta Physiol. Scand., 56, Suppl. 196:1-28.

Carlsson, A., Magnusson, T., and Rosengren, E. (1963). "5-Hydroxytryptamine of the spinal cord normally and after transection," Experientia, 19:359.

Carlsson, A., Rosengren, E., Bertler, A., and Nilsson, J. (1957). "Effect of reserpine on the metabolism of catecholamines," in: Psychotropic Drugs, S. Garattini and V. Chetti (editors), Elsevier, Amsterdam, pp. 363-372.

Carnicalli, A., Saba, P., Saba, G. C., and Marescotti, V. (1964). "The importance of the optic pathways in the control of the adrenal cortical activity," Folia Endocrinol. (Rome), 17:794-797.

Carr, L. A., and Moor, K. E. (1968). "Effects of reserpine and α-methyltyrosine on brain catecholamines and the pituitary-adrenal response to stress," Neuroendocrinology, 3:285-302.

Carr, L. A., and Moore, K. E. (1969). "Distribution and metabolism of norepinephrine after its administration into the cerebroventricular system of the cat," Biochem. Pharmacol., 18:1907-1918.

Carr, L. A., and Moore, K. E. (1970). "Effects of amphetamine on the contents of norepinephrine and its metabolites in the effluent of perfused cerebral ventricles of the cat," Biochem. Pharmacol., 19:2361-2374.

Casentini, S., De Poll, A., and Martini, L. (1957). "Hypothalamo-neurohypophyseal involvement in the corticotrophic action of acetylcholine," Brit. J. Pharmacol., 12:166-170.

Cerletty, A. (1958). "Biochemie, Physiologie und Pharmacologie des 5-Hydroxytryptamins (Serotonin, Enteramin)," Helvetica Medica Acta, 25:330-350.

Chatterjee, A. (1965). "Adrenal cortical response to administration of chlorpromazine in rats," Naturwissenschaften, 52:456.

Chernov, G. A. (1960). "Serotonin," Med. Radiologiya, No. 6:75-78.

Chowers, I., Hammel, H. T., Stromme, S. B., and McCann, S. M. (1964). "Comparison of effect of environmental and preoptic cooling on plasma cortisol levels," Amer. J. Physiol., 207:577-580.

Christina, J. J. (1950). "The adreno-pituitary system and population cycles in mammals," J. Mammal., 31:247-259.

Christian, J. J. (1959). "Lack of correlation between adrenal weight and injury in grouped male albino mice," Proc. Soc. Exptl. Biol. (New York), 101:166-168.

Christian, J. J. (1960). "Adrenocortical and gonadal responses of female mice to increased population density," Proc. Soc. Exptl. Biol. (New York), 104:330-332.

Christian, J. J. (1963). "Endocrine adaptive mechanisms and the physiologic regulation of population growth," in: Physiological Mammology, Academic Press, New York and London, Vol. 1, pp. 189-353.

Christian, J. J., Lloyd, J. A., and Davis, D. E. (1965). "The role of endocrines in the self-regulation of mammalian populations," Rec Progr. Hormone Res., 21: 501-571.

Christy, N. P., Knight, M., Longson, D., and Jailer, J. M. (1956). "Inhibition by chlorpromazine of the adrenocortical response to insulin hypoglycemia," J. Clin. Endocrinol. Metabol., 16:913-914.

Clark, C. T., Weissbach, H., and Udenfriend, S. (1954). "5-Hydroxytryptophan decarboxylase: preparation and properties," J. Biol. Chem., 210:139-148.

Clement, F., Ceccarelli, B., Cerati, E., Demonte, M., Felici, M., Motta, M., and Pecile, A. (1970). "Subcellular localization of neurotransmitters and releasing factors in the rat median eminence," J. Endocr., 48:205-213.

Colfer, H. F., de Groot, J., and Harris, G. M. (1950). "Pituitary gland and blood lymphocytes," J Physiol. (London), 111:328-334.

Compani, M., and Peruzzo, L. (1953). "Comportement des 17-cétostéroides neutres urinaires sous l'action des gangliop1égiques," Anesth. Analg., 10:378-381.

Consolo, S., Garattini, S., Ghielmetti, R., Morselli, P., and Valzelli, L. (1965). "The hydroxylation of tryptophan *in vitro* by brain," Life Sci., 4:625-630.

Corrodi, H., Fuxe, K., and Hökfelt, T. (1968). "The effect of immobilisation stress on the activity of central monoamine neurons," Life Sci., 7:107-112.

Costa, E. (1960). "The role of serotonin in neurobiology," Internat. Rev. Neurobiol., 2:175-221.

Craigie, E. H. (1925). An Introduction to the Finer Anatomy of the Central Nervous System Based upon That of the Albino Rat, University Press, Toronto.

Crane, G. E., and Wolfman, M. (1960). "Studies on the urinary excretion of 5-hydroxyindoleacetic acid and 17-hydroxycorticosteroids in patients treated with marsilid," J. Nervous Mental Diseases, 130:134-140.

Crawford, T. B. (1958). "The distribution of 5-hydroxytryptamine in the central nervous system of the dog," in: 5-Hydroxytryptamine, G. P. Lewis (editor), Pergamon Press, Oxford, pp. 20-26.

Crawford, J. M., and Curtis, D. R. (1964). "The excitation and depression of mammalian cortical neurons by amino acids," Brit. J. Pharmacol. and Chemotherapy, 23:313-329.

Crossland, J. (1953). "The significance of brain acetylcholine," J. Ment. Sci., 99:247-251.

Csordas, E., Endröczi, E., and Lissak, K. (1952). "Studies on the changes of the blood-lymphocyte count," Acta Physiol. Acad. Sci. Hung., 3:75-77.

Curtis, D. R. (1961). "The effect of drugs and amino acids upon neurons," in: Regional Neurochemistry, S. S. Kety and J. Elkes (editors), Pergamon Press, New York, pp. 403-421.

Curtis, D. R., and Eccles, R. M. (1958). "The excitation of Renshaw cells by pharmacological agents applied electrophoretically," J. Physiol. (London), 141: 435-445.

Cushman, P., Alter, S., and Hilton, I. G. (1966). "Cortisol secretion by the dog adrenal: effects of cyclic adenosine monophosphate, dichloroisoproterenol, dihydroergotamine and adrenaline," J. Endrocrinol., 34:271-272.

Dahlström, A. (1967). "The intraneuronal distribution of noradrenaline and the transport and life-span of amine storage granules in the sympathetic adrenergic neuron," Arch. Pharmakol. Exptl. Pathol., 257:93-115.

Dahlström, A., and Fuxe, K. (1965a). "Evidence for the existence of monoamine-containing neurons in the central nervous system. I. Demonstrations of monoamines in the cell bodies of brain stem neurons," Acta Physiol. Scand., 62, Suppl. 232:1-55.

Dahlström, A., and Fuxe, K. (1965b). "Evidence for the existence of monoamine neurones in the central nervous system. II. Experimentally induced changes in the intraneuronal amine levels of bulbospinal nervous system," Acta Physiol. Scand., 64, Suppl. 247:1-36.

Daily, W. J. R., and Ganong, W. F. (1958). "The effect of ventral hypothalamic lesions on sodium and potassium metabolism in the dog," Endocrinology, 62:442-454.

Danford, H. G., and Danford, P. A. (1951). "Circulating leukocytes in sodium chloride deficient rats and their response to epinephrine, ACTH, cortisone, and deoxycorticosterone acetate," Amer. J. Physiol., 166:514-527.

D'Angelo, S. A., Snyder, I., and Grodin, I. M. (1964). "Electrical stimulation of the hypothalamus: simultaneous effects on the pituitary-adrenal and thyroid systems of the rat," Endocrinology, 75:417-427.

D'Angelo, S. A., and Young, R. (1966). "Chronic lesions and ACTH: effects of thyroid hormones and electrical stimulation," Amer. J. Physiol., 210:795-800.

Danilyak, I. G. (1962). "Thorn's tests in patients with bronchial asthma," Clin. Med., No. 4:64-66.

Davis, D. E., and Christian, J. J. (1957). "Relations of adrenal weight to social rank of mice," Proc. Soc. Exptl. Biol. (New York), 94:728-731.

Davison, A. N. (1958). "Physiological role of monoamine oxidase," Physiol. Rev., 38:729-745.

De Groot, I., and Harris, G. W. (1950). "Hypothalamic control of the anterior pituitary gland and blood lymphocytes," J. Physiol. (London), 111:335-346.

De Maio, D. (1959a). "Influence of adrenalectomy and hypophysectomy on cerebral serotonin," Science, 129:1678-1679.

De Maio, D. (1959b). "Contributo alla conescenza del meccanismo d'azione della beta-feniletilidrazine (Fenelzina)," Acta Neurol., 14:761-777.

De Maio, D., and Marobbio, C. (1961). Influenza ditaluni ormoni sul contenuto in serotonina del cervello di ratto (normale e surrenectomizzato), "Arch. Sci. Med., 111:369-373.

De Robertis, E. (1971). "Molecular biology of synaptic receptors," Science, 171:963-971.

De Schaepdryver, A. F., and Preziosi, P. (1959). "Iproniazide et effects pharmacologiques sur la médullo-cortico-surrénale," Arch. Internat. Pharmacodyn. et Therap., 119:506-510.

De Schaepdryver, A., Preziosi, P., and Scapagnini, U. (1969). "Brain monoamines and adrenocortical activation," Brit. J. Pharmacol., 35:460-467.

Dingell, J. V., Owens, M. L., Norvich, M. R., and Sulser, F. (1967). "On the role of norepinephrine biosynthesis in the central action of amphetamine," Life Sci., 6:1155-1162.

Dixit, B. N., and Buckley, J. P. (1969). "Brain 5-hydroxytryptamine and their pituitary activation by stress," Neuroendocrinology, 4:32-41.

Dordoni, F., and Fortier, C. (1950). "Effect of eserine and atropine on ACTH release," Proc. Soc. Exptl. Biol. (New York), 75:815-816.

Eccles, J. C., Fatt, P., and Koketsu, K. (1954). "Cholinergic and inhibitory synapses in a pathway from motor-axon collaterals to motoneurons," J. Physiol. (London), 126:524-562.

Eechaute, W., Lacroix, E., and Coessens, R. (1962a). "L'activité corticosurrénalienne sous l'influence de la réserpine et de l'iproniazide," Arch. Internat. Physiol. Biochem., 70:117-119.

Eechaute, W., Lacroix, E., Leusen, I., and Bouckaert, J. J. (1962b). "L'activité du cortex surrénalien sous l'influence de la réserpine et de l'iproniazide," Arch. Internat. Pharmacodyn. Therap., 139:403-414.

Egdahl, R. H. (1959). "Adrenal cortical function in untraumatized dogs, before and after removal of higher central nervous system structures," Clin. Res., 7:252.

Egdahl, R. H. (1960a). "Adrenal cortical and medullary responses to trauma in dogs with isolated pituitaries," Endocrinology, 66:200-216.

Egdahl, R. H. (1960b). "The effect of brain removal, decortication and midbrain transsection on adrenal cortical function in dogs," Acta Endocrinol., 35, Suppl. 51:49-50.

Egdahl, R. H. (1961a). Corticosteroid secretion following caval constriction in dogs with isolated pituitaries," Endocrinology, 68:226-231.

Egdahl, R. H. (1961b). "Cerebral cortical inhibition of pituitary-adrenal secretion," Endocrinology, 68:574-581.

Egdahl, R. H. (1962). "Further studies on adrenal cortical function in dogs with isolated pituitaries," Endocrinology, 71:926-935.

Egdahl, R. H., and Richards, J. B. (1956). "Effect of chlorpromazine on pituitary ACTH secretion in the dog," Amer. J. Physiol., 185:235-238.

Eleftheriou, B. E., Zolovick, A. J., and Pearsed, F. (1966). "Effect of amygdaloid lesions on pituitary-adrenal axis in the doormouse," Proc. Soc. Exptl. Biol. (New York), 122:1259-1262.

Elliott, K. A. C., and Jasper, H. H. (1959). "Gamma-aminobutyric acid," Physiol. Rev., 39:383-405.

Elliott, K. A. C., Swank, R. L., and Henderson, N. (1950). "Effects of anesthetics and convulsants on acetylcholine content of brain," Amer. J. Physiol., 162:469-474.

Ely, R. S., Bray, P. F., Raile, R. B., and Kelley, V. C. (1954). "Studies of 17-hydroxycorticosteroids. V. Responses of 17-hydroxycorticosteroids, eosinophils, and glucose to ACTH and epinephrine," J. Clin. Investig., 33:1587-1593.

Endröczi, E. (1961). "Contributions to the hypothalamic control of pituitary, ovarian and adrenal cortical function," Acta Physiol. Acad. Sci. Hung., 18:301-307.

Endröczi, E., Bata, G., and Martin, J. (1958a). "Untersuchungen über die Sekretion von Nebennierenrindenhormonen," Endokrinologie, 35:280-290.

Endröczi, E., Kovacs, S., and Bohus (1958b). "Changes in behavior and endocrine activation in response to the stimulation of orbito-frontal cortical areas," Acta Physiol. Acad. Sci. Hung., 14:39-44.

Endröczi, E., Kovacs, S., and Lissak, K. (1956). "Die Wirkung der Hypothalamusreizung auf das endokrine und somatische Verhalten," Endrokrinologie, 33:271-278.

Endröczi, E., and Lissak, K. (1960). "The role of the mesencephalon, diencephalon and archicortex in the activation and inhibition of the pituitary-adrenocortical system," Acta Physiol. Acad. Sci. Hung., 17:39-55.

Endröczi, E., and Lissak, K. (1962). "Interrelations between paleocortical activity and pituitary-adrenocortical function," Acta Physiol. Acad. Sci. Hung., 21: 257-263.

Endröczi, E., and Lissak, K. (1963). "Effect of hypothalamic and brain stem structure stimulation on pituitary adrenocortical function," Acta Physiol. Acad. Sci. Hung., 24:67-77.

Endröczi, E., Lissak, K., Bohus, B , and Kovacs, S. (1959). "The inhibitory influence of archicortical structures on pituitary-adrenal function," Acta Physiol. Acad. Sci. Hung., 16:17-22.

Endröczi, E., and Nagy, D. (1951). "Contributions of the mechanism of the lymphopenia caused by the adrenal cortex," Acta Physiol. Acad. Sci. Hung., 2:11-15.

Endröczi, E., Schreiberg, G., and Lissak, K. (1963). "The role of central nervous activating and inhibitory structures in the control of pituitary-adrenocortical function. Effect of intracerebral cholinergic and adrenergic stimulation," Acta Physiol. Acad. Sci. Hung., 24:211-221.

Enerbäck, L. (1960). "Influence of adrenalectomy on urinary excretion of 5-hydroxyindole acetic acid in the rat," Endocrinology, 67:717-719.

Erspamer, V. (1966). "Peripheral physiological and pharmacological actions of indolealkylamines," in: Handbook of Experimental Pharmacology (New series), 5-Hydroxytryptamine and Related Indolealkylamines, Springer, Berlin, 19:270-271.

Erspamer, V., and Bertaccini, G. (1962). "Observations on the antidiuretic action and the fate of 5-hydroxy-DL-tryptophan in the rat organism," Arch. Internat. Pharmacodyn. Therap., 137:6-23.

Eskin, I. A. (1960). "Some problems in the physiology of the pituitary and adrenal cortex," in: Current Problems in Endocrinology, No. 1 [in Russian], pp. 169-190.

Eskin, I. A., and Mikhailova, N. V. (1960). "The functional state of the pituitary and adrenal cortex in young and old rats in a situation of stress," Probl. Endokrinol. i Gormonoter., 6(3):3-8.

Eskin, I. A., and Shchedrina, R. N. (1964). "New data on the hypothalamic regulation of the adrenocorticotrophic function of the pituitary," in: Pituitary-Adrenal Cortex [in Russian], Naukova Dumka, Kiev, pp. 18-28.

Eskin, I. A., and Shchedrina, R. N. (1966). "Role of the catecholamines of the central nervous system in the hypothalamic regulation of the adrenocorticotrophic function of the pituitary," in: Physiology and Pathology of the Hypothalamus [in Russian], Nauka, Moscow, pp. 147-150.

Eskin, I. A., and Shchedrina, R. N. (1968). "Catecholamines of the hypothalamus and reaction of the organism to stressors," Abstracts of Proceedings of the 5th All-Union Conference on Neurochemistry [in Russian], Tbilisi, pp. 224-225.

Eskin, I. A., Skobel'skaya, Yu. B., and Mikhailova, N V. (1957). "Role of the nervous system in the formation and mechanism of action of pituitary hormones," Abstracts of Proceedings of a Conference on the Mechanism of Action of Hormones [in Russian], Kiev, pp. 162-163.

Etkin, W. (1967). "Social behaviour from fish to man," Phoenix Books, Chicago, London.

Euler, U. S. (1964). "Quantitation of stress by catecholamine analysis," Clin. Pharmacol. Therap., 5:398-404.

Euler, U. S., and Folkow, B. (1958). "The effect of stimulation of autonomic areas in the cerebral cortex upon the adrenaline and noradrenaline secretion from the adrenal gland in the cat," Acta Physiol. Scand., 42:313-320.

Euler, U. S., Gemzell, C. A., Levi, L., and Ström, G. (1959). "Cortical and medullary adrenal activity in emotional stress," Acta Endocrinol., 30:567-573.

Farrel, G. L., and McCann, S. M. (1952). "Detectable amounts of adrenocorticotrophic hormone in blood following epinephrine," Endocrinology, 50:274-276.

Feldberg, W. (1954). "Synthesis of acetylcholine by tissues of the central nervous system," J. Physiol. (London), 193:367-402.

Feldberg, W. (1966). "Introductory remarks," Pharmacol. Rev., 18:713-716.

Feldman, S., Conforti, N., Chowers, I., and Davidson, N. (1970). "Pituitary-adrenal activation in rats with medial basal hypothalamic islands," Acta Endrocrinol., 63:405-414.

Feldman, S., Conforti, N., and Davidson, J. M. (1965/1966). "Adrenocortical responses in rats with corticosteroid and reserpine implants," Neuroendocrinology, 1: 228-239.

Fendler, K., Karmos, G., and Telegdy, G. (1961). "The effect of hippocampal lesion on pituitary-adrenal function," Acta Physiol. Acad. Sci. Hung., 20:293-297.

Ferguson, R. W., Folkow, B., Mitts, M. G., and Hoff, E. C. (1957). "Effect of cortical stimulation upon epinephrine activity," J. Neurophysiol., 20:329-339.

Filk, H., and Loeser, A. (1954). "Zur Frage der Gewöhnung bei Ratten nach oraler Megaphenzufuhr," Klin. Wschr., 32:661-662.

Fiore-Donati, L., Pollice, L., and Chieco-Bianchi, L. (1959). "Response of adrenal and preputial glands of rats to administration of 5-hydroxytryptamine," Experientia, 15:193-195.

Fiorica, V., and Muchl, S. (1962). "Relationship between plasma levels of 17-hydroxy-corticosteroids (17-OH-CS) and a psychological measure of manifest anxiety," Psychosom. Med., 24:596-599.

Fischer, P., Renson, J., and Ciccarone, P. (1959). "Effect de la 5-hydroxytryptamine sur la surrénale du rat normal, morphiné ou nouveau-né," Arch. Internat. Physiol. Biochem., 67:147.

Fleckenstein, A., and Burn, J. H. (1953). "The effect of denervation on the action of sympathomimetic amines on the nictitating membrane," Brit. J. Pharmacol., 8:69-78.

Fleckenstein, A., and Stöckle, D. (1955). "Zum Mechanismus der Wirkungsverstärkung und Wirkungsabschwächung sympathomimetischer Amine durch Cocain und andere Pharmaka. II. Die Hemmung der Neuro-Sympathomimetika durch Cocain," Arch. Exptl. Pathol. und Pharmakol., 224:401-415.

Florey, E. (1967). "Neurotransmitters and modulators in the animal kingdom," Federat. Proc. 26:1164-1178.

Folkow, B., and Euler, U. S. (1954). "Selective activation of noradrenaline and adrenaline producing cells in the suprarenal gland of the cat by hypothalamic stimulation," Circulation Res., 2:191-195.

Folkow, B., and Rubinstein, E. H. (1966). "Cardiovascular effects of acute and chronic stimulations of the hypothalamic defense area in the rat," Acta Physiol. Scand., 68:48-57.

Fortier, C. (1951a). "Dual control of adrenocorticotrophin release," Endocrinology, 49:782-788.

Fortier, C. (1951b). "Dual control of ACTH release," J. Clin. Endocrinol. and Metabol., 11:751.

Fortier, C. (1951c). "Studies on the adrenocorticotrophic effect of audiogenic stimulation," Rev. Canad. Biol., 10:67-68.

Fortier, C. (1962). "Adenohypophysis and adrenal cortex," Annual Rev. Physiol., 24:223.

Fortier, C., Harris, G. W., and McDonald, I. R. (1957). "The effect of pituitary stalk section on the adrenocortical response to stress in the rabbit," J. Physiol. (London), 136:344-363.

Foss, M. L., Barnard, R. J., and Tipton, C. M. (1971). "Free 11-hydroxycorticosteroid levels in working dogs as affected by exercise training," Endocrinology, 89: 96-104.

Franksson, C., and Gemzell, C. A. (1955). "Adrenocortical activity in the preoperative period," J. Clin. Endocrinol. Metabol., 15:1069-1072.

Franksson, C., Gemzell, C., and Euler, U. S. (1954). "Cortical and medullary adrenal activity in surgical and allied condition," J. Clin. Endocrinol. Metabol., 14:608-622.

Franschini, F., Mengili, G., Motta, M., and Martini, L. (1964). "Midbrain and feedback control of adrenocorticotrophin secretion," Endocrinology, 75:765-769.

Freedman, D. X., Barchas, J. D., and Schoenbrun, R. L. (1962). "Response of brain amines to exhaustion-stress and LSD-25," Federat. Proc., 21:337-337.

Fritz, J., and Levine, R. (1951). "Action of cortical steroids and norepinephrine on vascular response of stress in adrenalectomized rats," Amer. J. Physiol., 165: 456-465.

Fuxe, K. (1965). "Evidence for the existence of monoamine neurons in the central nervous system. IV. Distribution of monoamine nerve terminals in the central nervous system," Acta Physiol. Scand., 64, Suppl. 247:41-84.

Fuxe, K., Corrodi, H., Hökfelt, T., and Jonsson, G. (1970a). "Central monoamine neurons and pituitary-adrenal activity," in: Progress in Brain Research, D. de Wied and J. A. W. M. Weijnen (editors), Elsevier, Amsterdam, Vol. 32, pp. 42-56.

Fuxe, K., Hökfelt, T., and Jonsson, G. (1970b). "Participation of central monoaminergic neurons in the regulation of anterior pituitary secretion," in: Neurochemical Aspects of Hypothalamic Function, L. Martini and J. Meites (editors), Academic Press, New York and London, pp. 61-83

Gann, D. S., and Egdahl, R. H. (1965). "Responses of adrenal corticosteroid secretion to hypotension and hypovolemia," J. Clin. Investig., 44:1-7.

Ganong, W. F., and Betz, D. (1960). "The effect of chlorpromazine on pituitary-adrenal function in the dog," Acta Endocrinol., 35, Suppl. 51:45.

Ganong, W. F , and Hume, D. M. (1954). "Absence of stress-induced and 'compensatory' adrenal hypertrophy in dog with hypothalamic lesions," Endocrinology, 55:474-483.

Ganong, W. F., Nolan, A. M., Dowdy, A., and Luetscher, I. A. (1961). "The effect of hypothalamic lesions on adrenal secretion of cortisol, corticosterone, 11-desoxycortisol and aldosterone," Endocrinology, 68:169-171.

Garattini, S., Gaiardoni, P., Mortari, A., and Palma, V. (1961). "Increased toxicity of serotonin in adrenalectomized animals," Nature, 190:540-541.

Garattini, S., Kato, R., and Valzelli, L. (1960). "Biochemical and pharmacological effects induced by electroshock," Psychol. Neurol. (Basel), 140:190-206.

Garattini, S., and Valzelli, L. (1957). "Serotonin and electroshock," in: Psvchotropic Drugs, W. Garattini and Ghetti (editors), Elsevier, Amsterdam, pp. 428-436.

Garattini, S., and Valzelli, L. (1965). Serotonin, Elsevier, Amsterdam.

Gatgounis, J. (1965). "Structural activity relationships of a series of amines injected before and after monoamine oxidase inhibitor," Arch. Internat. Pharmacodyn. et Therap., 154:412-420.

Gellhorn, E., and Frank, S. (1948). "Sensitivity of the lymphopenic reaction to adrenalin,' Proc. Soc. Exptl. Biol. (New York), 69:426-429.

Gellhorn, E., and Frank, S. (1949). "Lymphopenia and the secretion of adrenalin," Proc. Soc. Exptl. Biol. (New York), 71:112-115.

Gellhorn, E., and Loofbourrow, G. N. (1963). Emotions and Emotional Disorders [Russian translation], Mir, Moscow.

Georges, G. (1957). "Influence de la 5-hydroxytryptamine sur la reactivité du couple-hypophyso-surrénalien chez le rat," Compt. Rend. Soc. Biol., 151:692-695.

Georges, G., and Cahn, J. (1953). "Couple hypophyso-purrénalien et hibernation," Anesth. Analg., 10:409-419.

Georges, G., and Herold, M. (1958). "Influence de la sérotonine et du marsilid sur la réactivité de l'axe hypothalamo-hypophyso-surrénalien chez le rat (II)," Compt. Rend. Soc. Biol., 152:436-440.

Gershberg, H., Fry, E. G., Brobeck, J. R., and Long, C. N. N. (1950). "The role of epinephrine in the secretion of the adrenal cortex," Yale J. Bio. and Med., 23:32.

Gervazieva, I. D. (1962). "Character of the change in the ascorbic acid concentration in the adrenals during an increase in their function," Probl. Éndokrinol. i Gormonoter., 8(1):81-83.

Gey, K. F., and Pletscher, A. (1961). "Activity of monoamine oxidase in relation to the 5-hydroxytryptamine and norepinephrine content of the rat brain," J. Neurochem., 63:239-243.

Gibbons, J. L., and McHugh, P. R. (1962). "Plasma cortisol in depressive illness," Psychiatr. Res., 1:162-171.

Girod, C. (1961). "Action de la réserpine sur la corticosurrénale du singe Macacus sylvanus L.," Compt. Rend Soc. Biol. (Paris), 155:1628-1630.

Giuliani, G., Martini, L., and Pecile, A. (1961). "Midbrain section and release of ACTH following stress," Acta Neuroveget., 23:21-34.

Giuliani, G., Motta, M., and Martini, L. (1961). "Reserpine and corticotrophin secretion," Acta Endocrinol., 51:203-209.

Glin, J. H. (1957). Cortisone Therapy [Russian translation], Medgiz, Moscow.

Gloor, P. (1956). "Telencephalic influence upon the hypothalamus," in: Hypothalamic-Hypophyseal Interrelationships, W. S. Fields and R. Guillemin (editors), Charles C. Thomas, Springfield, Ill., pp. 74-114.

Glowinski, J., and Baldessarini, R. J. (1966). "Metabolism of norepinephrine in the central nervous system," Pharmacol. Rev., 18:1201-1238.

Glowinski, J., and Iversen, L. L. (1966a). "Regional studies of catecholamines in the rat brain. I. The disposition of [^{3}H]-norepinephrine, [^{3}H]-dopamine and [^{3}H]-dopa in various regions of the brain," J. Neurochem., 13:655-669.

Glowinski, J., and Iversen, L. L. (1966b). "Regional studies of catecholamines in the rat brain. III. Subcellular distribution of endogenous and exogenous catecholamines in various brain regions,' Biochem. Pharmacol., 15:977-987.

Godowski, Z. Z. (1948). "Eosinopenia caused by adrenaline infusion and by insulin hypoglycaemia," Brit. Med. J., 1:46-48.

Goldberg, M. E., and Salama, A. I. (1969). "Amphetamine toxicity and brain monoamines in three models of stress," Toxicol. Appl. Pharmacol., 14:447-456.

Goldfien, A., and Ganong, W. F. (1962). "Adrenal medullary and adrenal cortical response to stimulation of diencephalon," Amer. J. Physiol., 202:205-211.

Gordon, M. L. (1950). "An immediate response of the demudullated adrenal gland to stress," Endocrinology, 47:13-18.

Gordon, P. (1961). "Changes in blood serotonin during exposure to cold," Nature, 191:183-184.

Gordon, R., Spector, S., Sjoerdsma, A., and Udenfriend, S. (1966). "Increased synthesis of norepinephrine and epinephrine in the intact rat during exercise. and exposure to cold," J. Pharmacol. Exptl. Therap., 153:440-447.

Grashchenkov, N. I. (1963). "The hypothalamic region. Anatomy, physiology, and pathology," in: The Physiology and Pathology of the Diencephalic Region of the Brain [in Russian], Izd. An SSSR, Moscow, pp. 5-82.

Green, H., and Sawyer, J. L. (1966). "Demonstration, characterization, and assay procedure of tryptophan hydroxylase in rat brain," Analyt. Biochem., 15:453-464.

Green, J. D. (1956). "Neural pathways to the hypophysis," in: Hypothalamic-Hypophyseal Interrelationships, W. S. Fields and R. Guillemin (editors), Charles C. Thomas, Springfield, Ill., pp. 3-14.

Green, J. D. (1958). "The olfactory brain, its relation to behavior and its connection with the reticular activating system," in: The Reticular Formation of the Brain [Russian translation], Medgiz, Moscow (1962), pp. 533-543.

Green, I. P., Paasonen, M. K., and Giarman, N. I. (1957). "Blood 5-hydroxytryptamine (serotonin) levels after reserpine and electroshock therapy," Proc. Soc. Exptl. Biol. (New York), 94:428-430.

Gromova, E. A. (1966). Serotonin and Its Role in the Body [in Russian], Meditsina, Moscow.

Guillemin, R. (1955). "A re-evaluation of acetylcholine, adrenaline, noradrenaline and histamine as possible mediators of the pituitary adrenocorticotrophic activation by stress," Endocrinology, 56:248-255.

Guillemin, R. (1956). "Hypothalamic-hypophyseal interrelationships in the production of pituitary hormones *in vitro*," in: Hypothalamic-Hypophyseal Interrelationships, W. S. Fields and R. Guillemin (editors), Charles C. Thomas, Springfield, Ill., pp. 46-57.

Guillemin, R. (1957). "Central action drugs and pituitary-adrenal responses to stress," in: Brain Mechanisms and Drug Action, Springfield, Ill., pp. 99-109.

Guillemin, R. (1962). "Sur la nature des substances hypothalamiques qui contrôlent la secretion des hormones ante hypophysaires," J. Physiol. (Paris), 55:7-44.

Guillemin, R., Hearn, W. R., Cheek, W. R., and Housholder, D. E. (1957). "Control of corticotrophin release: further studies with *in vivo* methods," Endocrinology, 60:488-506.

Guillemin, R., and Rosenberg, B. (1955). "Humoral hypothalamic control of anterior pituitary. A study with combined tissue cultures," Endocrinology, 57:599-607.

Guillemin, R., Schally, A. K., Lipscomb, H. S., Andersen, R. N., and Long, J. M. (1962). "On the presence in hog hypothalamus of β-corticotrophin releasing factor, α- and β-melanocyte stimulating hormone, adrenocorticotropin, lysine-vasopressin and oxytocin," Endocrinology, 70:471-477.

Gumulka, W., Samanin, R., Garattini, S., and Valzelli, L. (1969), "Effect of stimulation of midbrain raphe on serotonin (5-HT) level and turnover in different areas of rat brain," Europ. J. Pharmacol., 8:380-384.

Halász, B. (1969). "The endocrine effects of isolation of the hypothalamus from the rest of the brain," in: Frontiers in Neuroendocrinology, W. F. Ganong and L. Martini (editors), pp. 307-342.

Halász, B., and Pupp, L. (1965). "Hormone secretion of the anterior pituitary gland after physical interruption of all nervous pathways to the hypophyseotrophic area," Endocrinology, 77:553-562.

Halász, B., Slusher, M. A., and Gorski, R. A. (1967a). "Adrenocorticotrophic hormone secretion in rats after partial or total deafferentation of the medial basal hypothalamus," Neuroendocrinology, 2:43-55.

Halász, B., Vernikos-Danellis, J., and Gorski, R. A. (1967b). "Pituitary ACTH content in rats after partial or total interruption of neural afferents to the medial basal hypothalamus," Endocrinology, 81:921-924.

Halberg, F. (1954). "Eosinopenic effects of tryptamines in mice. Synergism of effects of cortisone and serotonin," Amer. J. Physiol., 179:309-313.

Hale, H. B. (1952). "Adrenocortical activity associated with exposure to low-frequency sounds,' Amer. J. Physiol., 171:732.

Hamilton, L. (1952). "Effect of dibenzyline (an adrenolytic agent) on epinephrine-induced leukocyte changes in dogs," Amer. J. Physiol., 171:733.

Hamilton, L. H., and Horvath, S. M. (1953). "Effect of dibenzyline on prolonged leukocyte response following epinephrine administration," Amer. J. Physiol., 172:725.

Handlon, J. H., Wadeson, R. W., Fishman, R., Sachar, E. J., Hamburg, D. A., and Mason, J. W. (1962). "Psychological factors lowering plasma 17-hydroxycorticosteroid concentration," Psychosom. Med., 24:535-542.

Haning, R., Tait, S. A. S., and Tait, J. F. (1970). "*In vitro* effects of ACTH, angiotensins, serotonin and potassium on steroid output and conversion of corticosterone to aldosterone by isolated adrenal cells," Endocrinology, 87:1147-1152.

Harris, G. W. (1955). "Neural Control of the Pituitary Gland, Arnold, London.

Harris, G. W. (1956). "Hypothalamic control of the anterior lobe of the hypophysis," in: Hypothalamic-Hypophyseal Interrelationships, W. S Fields and R. Guillemin (editors), Charles C. Thomas, Springfield, Ill., pp. 31-42.

Harris, G. W. (1958). "The reticular formation, the stress response, and endocrine activity," in: The Reticular Formation of the Brain [Russian translation], Medgiz, Moscow (1962), pp. 191-203.

Harwood, C. T , and Mason, J. W. (1956). "Effects of intravenous infusion of autonomic agent on peripheral blood 17-hydroxycorticosteroid levels in the dog," Amer. J. Physiol., 186:445-452.

Harwood, C. T., and Mason, J. W. (1957). "Acute effects of tranquilizing drugs on the anterior pituitary-ACTH mechanism," Endrocrinology, 60:239-246.

Hassler, C. R., and McCook, R. D. (1971). "Hypothalamic blood flow in cats during injection of biogenic amines," Am. J. Physiol., 220:196-201.

Hebb, C. O., and Whittaker, V. P. (1958). "Intracellular distribution of acetylcholine and choline acetylase," J. Physiol. (London), 142:187-196.

Heller, A., and Moore, R. (1965). "Effect of central nervous system lesions on brain monoamines in the rat," J. Pharmacol. Exptl. Therap., 150:1-9.

Henkin, R. I., and Knigge, K. M. (1963). "Effect of sound on the hypothalamic-pituitary-adrenal axis," Amer. J. Physiol., 204:710-714.

Herz, A., and Nácimiento, A. C. (1965). "Über die Wirkung von Pharmaka auf Neurone des Hippocampus nach mikroelektrophoretischer Verabfolgung," Arch. Exptl. Pathol. und Pharmakol., 251:295-314.

Hess, S. M., Shore, P. A., and Brodie, B. B. (1956). "Persistence of reserpine action after the disappearance of drug from brain: effect on serotonin," J. Pharmacol. Exptl. Therap., 118:84-89.

Hickey, M. C., and Barnes, C. D. (1971). "Spinal cord effects of a brain stem cholinergic system," Proc. Western Pharmacol., Soc., 14:154-160.

Hicks, R., and West, G. B. (1958a). "Adrenal cortical hormones and the formation of histamine and 5-hydroxytryptamine," Nature, 181:1342, 1343.

Hicks, R., and West, G. B. (1958b). "Adrenalectomy and tissue amines," Nature, 182: 401-402.

Higginbotham, R. D. (1962). "Influence of adrenalectomy and cortisol on resistance of mice to histamine, serotonine, anaphylactic and endotoxin shocks," J. Allergy, 33:35-44.

Hill, S. R., Goetz, F. C., Fox, H. M., Murawski, B. J., Gray, S. J., Krakauer, L. J., and Thorn, G. W. (1955). "Adrenocortical and psychologic response to stress in man," J. Clin. Endocrinol. Metabol., 15:887.

Hill, S. R., Goetz, F. C., Fox, H. M., Murawski, B. J., Krakauer, L. J., Reifenstein, R. W., Gray, S. J., Reddy, W. J., Hedberg, S. E., Sy. Mark, J. R., and Thorn, G. W. (1956). "Studies on adrenocortical and psychological response to stress in man," A. M. A. Arch. Intern. Med., 97:269-298.

Hillarp, N. A., Fuxe, K., and Dahlström, A. (1966a). "Central monoamine neurons," in: Mechanisms of Release of Biogenic Amines. Proceedings of the International Wenner-Gren Symposium held in Stockholm, 1965, Pergamon Press, Oxford, pp. 31-36.

Hillarp, N. A., Fuxe, K., and Dahlström, A. (1966b). "Demonstration and mapping of central neurons containing dopamine, noradrenaline, and 5-hydroxytryptamine and their reactions to psychopharmaca," Pharmacol. Rev., 18:727-741.

Hirai, K. (1966). "Effect of stimulation of the cerebellum on noradrenalin and adrenaline secretion of the adrenal gland in the cat," Tohoku J. Exptl. Med., 88:215-220.

Hirsch, G. H., and Moor, K. E. (1968). "Brain catecholamines and the reserpine-induced stimulation of the pituitary-adrenal system," Neuroendocrinology, 3:398-405.

Holman, R. B., and Vogt, M. (1970). "Release of 5-hydroxytryptamine (5-HT) from septum and caudate nucleus," J. Physiol., 210:163.

Holtz, P., Balter, H., Westermann, E., and Wezber, E. (1957). "Beeinflussung der Evipannarkose durch Reserpine, Iproniazid und Biogene Amine," Arch. Exptl. Pathol. und Pharmakol., 231:333-348.

Holzbauer, M., and Vogt, M. (1954). "The action of chlorpromazine on diencephalic sympathetic activity and on the release of adrenocorticotrophic hormone," Brit. J. Pharmacol., 9:402-407.

Hornikiewicz, O. (1966). "Dopamine (3-hydroxytyramine) and brain function," Pharmacol. Rev., 18:925-964.

Hortling, H., and Pekkarinen, A. (1949). "Lymphocytopenia and eosinopenia after prolonged intravenous adrenaline injection on man," Acta Endrocrinol., 2:356-364.

Horton, E. W. (1969). "Hypotheses on physiological roles of prostaglandins," Physiol. Rev., 49:122-161.

Hume, D. M. (1949). "The role of the hypothalamus in the pituitary-adrenal cortical response to stress," J. Clin. Investig., 28:790.

Hume, D. M. (1953). "The neuro-endocrine response to injury: present status of the problem," Ann. Surg., 138:548-557.

Hume, D. M. (1958). "The secretion of epinephrine, norepinephrine, and corticosteroids in the adrenal venous blood of the dog following single and repeated trauma," Surg. Forum, 8:111-115.

Hume, D. M. (1958). "The localization of hypothalamic control over various endocrine functions," in: The Reticular Formation of the Brain [Russian translation], Medgiz, Moscow (1962), pp. 211-225.

Hume, D. M., and Egdahl, R. H. (1959). "The importance of the brain in the endocrine response to injury," Ann. Surg., 150:697-712.

Hume, D. M., and Jackson, B. T. (1959). "Adrenal output of corticosteroids in the first 14 days after hypothalamic destruction in the dog," Federat. Proc., 18:481.

Hume, D. M., and Nelson, H. (1955). "Effect of hypothalamic lesions on blood ACTH levels and 17-hydroxycorticosteroid secretion following trauma in the dog," J. Clin. Endocrinol. Metabol., 15:839-840.

Hume, D. M., and Wittenstein, G. J. (1950). "The relationship of the hypothalamus to pituitary-adrenocortical function," in: Proceedings of the 1st Clinical ACTH Conference, I. R. Mote (editor), Blakiston, Philadelphia, pp. 134-145.

Ibayashi, H., Uchikawa, T., Motohashi, K., Fujita, T., Yoshida, Sh., Ohsawa, N., Muracawa, Sh., Nacamura, M , and Okinaka, Sh. (1963). "Effect of electrical stimulation of the limbic system on pituitary adrenal function: anterior and posterior cingular gyrus," Endocrinology, 73:816-818.

Il'yuchenok, R. Yu. (1965). Neurohumoral Mechanisms of the Reticular Formation of the Brain Stem [in Russian], Nauka, Moscow.

Irwin, R. L., and Smith, H J. (1960). "III. Cholinesterase inhibition by galanthamine and lycoramine," Biochem. Pharmacol., 3:147-148.

Iversen, L. L., and Glowinski, J. (1966). "Regional studies of catecholamines in the rat brain. II. Rate of turnover of catecholamines in various brain regions," J. Neurochem., 13:671-682.

Iwamoto, T., and Sato, S. (1963). "Effects of chlorpromazine, azacyclonol and chlordiazepoxide on brain catecholamine content of stressed rats," Japan J. Pharmacol., 13:66-73.

Jakobson, T., and Hortling, H. (1954). "Studies on the endocrine regulation of the number of circulating eosinophil cells in the rat," Acta Endocrinol., 15:265-281.

Joseph, S. A , Knigge, K. M., and Voloschin, L. (1969). "Effect of isolation of the hypothalamo-pituitary unit in newborn guinea pigs," Neuroendocrinology, 4:42-50.

Jouan, P. (1967). "Propriétes adrénolgomérulotrophiques de la 5-hydroxytryptamine (serotonine)," Pathol.-Biol., 15:1145-1153.

Kamberi, I. A., Mical, R. S., and Porter, J. C. (1971). "Effect of anterior pituitary perfusion and intraventricular injection of catecholamines on FSH release," Endocrinology, 88:1003-1011.

Kark, R. M., and Muehrcke, R. C. (1952). "Does adrenaline eosinopenia reflect pituitary-adrenal function in man?" Lancet, 1:1189-1190.

Katsuki, S. (1961). "A central control mechanism of the adrenocortical function: with special reference to the functional regulation within the hypothalamus," Acta Neuroveget. (Vienna), 23:50-57.

Keller, H. H. (1972). "Depletion of cerebral monoamines by *p*-chlorophenylalanine in the cat," Experientia, 28:177-178.

Kelley, V. C., Ely, R. S., Raile, R. B., and Bray, P. F. (1952). "Comparison of eosinophil and circulating 17-hydroxycorticosteroid response to epinephrine and ACTH," Proc. Soc. Exptl. Biol. (New York), 81:611-614.

Kendall, J. W., Allen, C., and Greer, M. A. (1965). "ACTH secretion in midbrain-transected rats," Endocrinology, 77:1091-1096.

Kim, Ch., and Kim, C. U. (1961). "Effect of partial hippocampal resection on stress mechanism in rats," Amer. J. Physiol., 201:337-340.

King, A. B., and Thomas, J. A. (1968). "Effect of exogenous dopamine on rat adrenal ascorbic acid," J. Pharmacol. Exptl. Therap., 159:18-21.

Kinsell, L. W., Michaels, G. D., Friskey, R. W., Brown, F. R., and Keasling, I. E. (1956). "Pituitary-adrenocortical response to insulin, epinephrine and a non-specific stressor in normal and abnormal patients," J. Clin. Endocrinol. Metabol., 16:912.

Kirkpatrick, W. E., and Lomax, P. (1970). "Temperature changes following ionophoretic injection of acetylcholine into the rostral hypothalamus of the rat," Neuropharmacology, 9:195-202.

Kivalo, E., and Rinne, U. K. (1961). "Activity of the hypothalamo-neurohypophyseal neurosecretory system after iproniazid treatment," Ann. Med. Exptl. Fenn., 39:50-56.

Knigge, K. M. (1961). "Adrenocortical response to immobilization in rats with lesions in hippocampus and amygdala," Proc. Soc. Exptl. Biol. (New York), 108:18-24.

Knigge, K. M., and Hays, M. (1963). "Evidence of inhibitive role of hippocampus in neural regulation of ACTH release," Proc. Soc. Exptl. Biol. (New York), 114: 67-69.

Koe, B. K., and Weisman, A. (1966). "Marked depletion of brain serotonin by *p*-chlorophenylalanine," Federat. Proc., 25:452.

Kokka, N., Garcia, J. F., George, R., and Elliot, H. W. (1972). "Growth hormone and ACTH secretion: evidence for an inverse relationship in rats," Endocrinology, 90:735-743.

Kolpakov, M. G., Potekhin, K. G., and Fedenkov, V. I. (1962a). "The mechanism of the eosinopenic response in terminal states," Byull. Éksperim. Biol. i Med., 54(11):36-39.

Kolpakov, M. G., Fedenkov, V. I., and Shushpannikova, O. V. (1962b). "The pathogenesis of the eosinopenic response in terminal states (Communication I)," Probl. Éndokrinol. i Gormonoter., 8(4):19-23.

Konradi, G. P., and Mikhel'son, M. Ya. (1935). "Chemical factors of nervous excitation," Uspekhi Sovr. Biol., 4(2):171-201.

Kopin, I. J. (1964). "Storage and metabolism of catecholamines: the role of monoamine oxidase," Pharmacol. Rev., 16:179-191.

Kordon, C., and Glowinski, J. (1970). "Role of brain catecholamines in the control of anterior pituitary functions," in: Neurochemical Aspects of Hypothalamic Function, L. Martini and J. Meites (editors), Academic, Press, New York and London, pp. 85-100.

Kovach, A. G. B., Monos, E., Koltay, E., and Desrius, A. (1970). "Effect of hypothalamic stimulation on adrenal blood flow and glycocorticoid release prior to and after acute hypophysectomy," Acta Physiol. Acad. Sci. Hung., 38:205-216.

Krieger, D. T., and Krieger, H. P. (1965). "The effects of intrahypothalamic injection of drugs on ACTH release in the cat," in: Proc. 2nd Internat. Congr. Endocrinol., Amsterdam, pp. 640-645.

Krieger, D. T., and Wagman, I. H. (1961). "Hypothalamic lesions and adrenal function in the cat," Acta Endocrinol., 38:88-98.

Krieger, H. P., and Krieger, D. T. (1970). "Chemical stimulation of the brain: effect on adrenal corticoid release," Am. J. Physiol., 218:1632-1641.

Krylov, S. S. (1963). "Pharmacological analysis of the chemical sensitivity of the carotid bodies," in: The Pharmacology of Neurotrophic Drugs [in Russian], Medgiz, Leningrad, pp. 161-170.

Kucherenko, T. M. (1961). "Changes in higher nervous activity in dogs under the influence of pipradol," Zh. Vyssh. Nervn. Deyat., 11:1052-1058.

Kudryavtsev, G. V. (1959). "Value of the eosinopenic response for the assessment of adrenocortical function in certain internal diseases," Sov. Meditsina, (11):31-35.

Kuhlenbeck, H. (1954). The Human Diencephalon, Karger, Basel.

Kulinskii, V. I. (1961). Changes in Catecholamine Metabolism in Acute Experimental Radiation Sickness. Author's Abstract of Candidate's Dissertation, Khar'kov.

Kulinskii, V. I. (1965). "Phasic changes in the catecholamine concentration in the adrenals during excitation of the sympathico-adrenal system," Proceedings of the 1st Congress of Endocrinologists of the Ukrainian SSR [in Russian], Khar'kov, pp. 244-246.

Kulinskii, V. I. (1968). "Role of the catecholamines in radioresistance," Proceedings of a Scientific Conference on the Physiology, Biochemistry, and Pathology of the Endocrine System [in Russian], Zdorov'ya, Kiev, pp. 90-92.

Laquer, G. L., McCann, S. M., Schreiner, L. H., Rosenberg, E., Rioch, D. Mc., and Anderson, E. (1955). "Alterations of adrenal cortical and ovarian activity following hypothalamic lesions," Endocrinology, 57:44-54.

Last, J. H., Jordan, P. H., Pitesky, I., and Siegel, B. M. (1950). "Resistance to epinephrine stress in the dog," Proc. Soc. Exptl. Biol. (New York), 74:96-102.

Laverty, R., and Sharman, D. F. (1965). "Modification by drugs of the metabolism of 3,4-dihydroxyphenylethylamine in the brain," Brit. J. Pharmacol., 24:759-772.

Lengyel, L. (1964). "Neurovegetative system and circulating eosinophils," Acta Neuroveget. (Vienna), 27:219-228.

Leroy, J. G., and Van Der Schoot, J. B. (1962). "5-Hydroxytryptamine and noradrenaline levels in brain of mice after nialamide, Ro-4-1530 and Ro-4-2308," Arch. Internat. Pharmacodyn. et Therap., 140:374-379.

Levi, R., and Maynert, E. W. (1962). 'Effects of stress on brain norepinephrine," Federat. Proc., 21:336.

Lewis, G. P. (1958). "5-Hydroxytryptamine," J. Pharmacy Pharmacol., 10:529-539.

Levitt, E. F., Persky, H., and Brady, J. P. (1964). Hypnotic Induction of Anxiety, Charles C. Thomas, Springfield, Ill.

Liberman, S. S. (1958). "New preparations stimulating the central nervous system. Piridrol and meridine," Proceedings of the 7th All-Union Conference of Pharmacologists [in Russian], Khar'kov, p. 84.

Lissak, K., and Endröczi, E. (1964). "Neuroendocrine interrelationships and behavioural processes," in: Major Problems in Neuroendocrinology, E. Bajusz and G. Jasmin, (editors) Karger, Basel, pp. 1-16.

Lissak, K., and Endröczi, E. (1965). "The neuroendocrinology control of adaptation," in: Modern Trends in Physiological Sciences, Vol. 25, Pergamon Press, Oxford.

Long, C. N. H. (1947). "The relation of cholesterol and ascorbic acid to the secretion of the adrenal cortex," Rec. Progr. Horm. Res., 1:1-99.

Long, C. N. H. (1952). "Regulation of ACTH secretion," Rec. Progr. Horm. Res., 7:75-97.

Long, C. N. H. (1956). "Pituitary-adrenal relationship," Annual Rev. Physiol. 18:409-432.

Long, C. N. H., and Fry, E. G. (1945). "Effect of epinephrine on adrenal cholesterol and ascorbic acid," Proc. Soc. Exptl. Biol. (New York), 59:67-68.

Lorenzen, L. C., and Ganong, W. F. (1967). "Effect of drugs to α-ethyl-tryptamine on stress-induced ACTH secretion in the dog," Endocrinology, 80:889-892.

Louverens, B., and Smelik, P. G. (1953). "The effect of hexamethonium chloride on the eosinopenic response during stress in rats," Acta Endocrinol., 14:27-32.

Lovenberg, W., Weissbach, H., and Udenfriend, S. (1962). "Aromatic L-amino acid decarboxylase," J. Biol. Chem., 237:89-93.

Lundberg, O. (1960). "Cortico-hypothalamic connexions in the rabbit," Acta Physiol. Scand., 49, Suppl.:171.

Lundborg, P., and Waldeck, B. (1971). "On the mechanism of amphetamine induced release of reserpine-resistant 3H-noradrenaline and 3H-α-methylnoradrenalin," Acta Pharmacol. Toxicol., 30:339-347.

Macchi, I. A., and Scotch, D. S. (1961). "*In vitro* effect of acetylcholine on adrenocortical secretion of incubated rat adrenals," Proc. Soc. Exptl. Biol. (New York), 106:324-327.

Magoun, H. W., Ranson, S. W., and Hetherington, A. (1937). "The liberation of adrenalin and sympathin induced by stimulation of the hypothalamus," Amer. J. Physiol., 119:615-622.

Makara, G. B., Stark, E., and Palkovits, M. (1970). "Afferent pathways of stressful stimuli: corticotrophin release after hypothalamic deafferentation," J. Endocr., 47:411-416.

Malhotra, C. L., and Prasad, K. (1962). "Effect of chlorpromazine and reserpine on the central amine content of different areas of the central nervous system of the dog," Brit. J. Pharmacol., 18:595-599.

Malhotra, C. L., and Pundlink, R. C. (1959). "The effect of reserpine on the acetylcholine content of different areas of the central nervous system of the dog," Brit. J. Pharmacol., 14:46-47.

Malmejac, M. I., Chardon, G., and Gros, A. (1947). "Sur la lymphopénie consécutive a l'injection intraveineuse d'adrénaline," Bull. Acad. Méd., 130:492-493.

Mandell, A. I., Chapman, L. F., Rand, R. W., and Walter, R. D. (1963). "Plasma corticosteroids: changes in concentration after stimulation of hippocampus and amygdala," Science, 139:1212.

Mantegazzini, P. (1966). "Pharmacological actions of indolealkylamines and precursor amino acids on the central nervous system," in: Handbook of Experimental Pharmacology (New series), 5-Hydroxytryptamine and Related Indolealkylamines, Springer, Berlin, Vol. 19, pp. 442-471.

Marks, B. H., Hall, M. M., and Bhattacharya, A. N. (1970). "Psychopharmacological effects and pituitary-adrenal activity," in: Progress in Brain Research, D. De Wied and J. A. W. M. Weijnen (editors), Elsevier, Amsterdam, Vol. 32, pp. 57-69.

Martin, I., Endröczi, E., and Bata, G. (1958). "Effect of the removal of amygdalic nuclei on the secretion of adrenal cortical hormones," Acta Physiol. Acad. Sci. Hung., 14:131-134.

Mashkovskii, M. D., and Al'tshuler, R. A. (1962). "The pharmacological properties of galanthamine iodimethylate,' Farmakol. i Toksikol., 25(2):168-175.

Mashkovskii, M. D., and Il'yuchenok, R. Yu. (1961). "The effect of galanthamine on the central nervous system," Zh. Nevropat. i Psikhiat., 61(2):166-175.

Mason, J. W. (1956). "Some aspects of the central nervous system regulation of ACTH secretion," J. Clin. Endocrinol. Metabol., 16:914.

Mason, J. W. (1958). "Plasma 17-hydroxycorticosteroid response to hypothalamic stimulation in the conscious Rhesus monkey," Endocrinology, 63:403-412.

Mason, J W. (1959a). "Plasma 17-hydroxycorticosteroid levels during electrical stimulation of the amygdaloid complex in conscious monkeys," Amer. J. Physiol., 196:44-48.

Mason, J. W. (1959b). "Psychological influences on the pituitary-adrenal-cortical system," Rec. Progr. Horm. Res., 15:345-389.

Mason, J. W. (1968). "A review of psychoendocrine research on the pituitary-adrenal cortical system," Psychosom. Med., 30:576-607.

Mason, J. (1962). "The central nervous regulation of ACTH secretion," in: The Reticular Formation of the Brain [Russian translation], Medgiz, Moscow, pp. 565-580 (Original version, 1958).

Mason, J. W., Brady, J. V., and Sidman, M. (1957a). "Plasma 17-hydrocorticosteroid levels and conditioned behavior in the Rhesus monkey," Endocrinology, 60:741-752.

Mason, J. W., Harwood, T., Rosenthal, N. (1957b). "Influence of some environmental factors on plasma and urinary 17-hydroxycorticosteroid levels in the Rhesus monkey," Amer. J. Physiol., 190:429-433.

Mason, J. W., Nauta. W I. H.. Brady I. V., Robinson, I. A., and Thach, I. S. (1960). "Limbic system influences on the pituitary-adrenal cortical system," Psychosom. Med., 22:322.

Matheson, G. K., Branch, B. J., and Taylor, A. N. (1971). "Effect of amygdaloid stimulation on pituitary-adrenal activity in conscious cats," Brain Res., 32:151-167.

Matlina, É. Sh., Shirinyan, É. A., and Rakhmanova, T. B. (1968). "Catecholamines, their precursors and metabolites in the rat brain under normal conditions and in a state of stress," Abstracts of Proceedings of the 5th All-Union Conference on Neurochemistry [in Russian], Tbilisi, pp. 168-169.

Matsuda, K., Kendall, J. W., Duyck, C., and Greer, M. A. (1963). "Neural control of ACTH secretion: effect of acute decerebration in the rat," Endocrinology, 72:845-852.

Matussek, N. (1966). "Hirnamine und psychotrope Pharmaka," Umschau Wissenschaft Technik, (12):400-405.

Maynert, E. W., and Levi, R. (1964). 'Stress-induced release of brain norepinephrine and its inhibition by drugs," J. Pharmacol. Exptl. Therap., 143:90-95.

McCann, S. M. (1952). "Effect of hypothalamic lesions on adrenal cortical response to stress," Amer. J. Physiol., 171:746.

McCann, S. M. (1953). "Effect of hypothalamic lesions on the adrenal cortical response to stress in the rat," Amer. J. Physiol., 175:13-20.

McCann, S M., and Dhariwal, A. P. S. (1966). "Hypothalamic releasing factors and the neurovascular link between the brain and the anterior pituitary," in: Neuroendocrinology, L. Martini and W. F. Ganong (editors), Academic Press, New York and London, Vol. 1, pp. 261-296.

McCann, S. M., and Haberland, P. (1960). "Further studies on the regulation of pituitary ACTH in rats with hypothalamic lesions," Endocrinology, 66:217-221.

McDermott, W. V., Fry, E. G., Brobeck, J. R., and Long, C. N. H. (1950a). "Release of adrenocorticotrophic hormone by direct application of epinephrine to pituitary grafts," Proc. Soc. Exptl. Biol. (New York), 73:609-610.

McDermott, W. V., Fry, E. G., Brobeck, J. R., and Long, C. N. H. (1950b). "The mechanism of control of adrenocorticotrophic hormone," Yale J. Biol. Med., 23:52-66.

McGeer, P. L. (1963). "Central amines and extrapyramidal function," Neuropsychiatry, 44:247-250.

McHugh, P. R., Black, W. C., and Mason, J. W. (1966). "Some hormonal responses to electrical self-stimulation in the *Macaca mutatta*," Amer. J. Physiol., 210:109-113.

McIntosh, F. O., and Oborin, P. E. (1953). "Release of acetylcholine from intact cerebral cortex," in: XIX Internat. Physiol. Congress. Abstr. Commun., Montreal, pp. 580-581.

McIsaac, W. M., and Page, J. H. (1959). "The metabolism of serotonin (5-hydroxytryptamine)," J. Biol. Chem., 234:858-864.

McKennee, C. T., Timiras, P. S., and Quay, W. B. (1965/1966). "Concentrations of 5-hydroxytryptamine in rat brain and pineal after adrenalectomy and cortisol administration," Neuroendocrinology, 1:251-256.

McLean, P. D. (1952). "Some psychiatric implications of physiological studies on frontotemporal portion of limbic system (visceral brain)," Electroenceph. Clin. Neurophysiol., 4:407-418.

Medakovic, M. (1959). "5-Hydroxytryptamine in blood during experimental traumatic shock in rats," Nature, 183:1685.

Mekhedova, A. Ya. (1961). "The effect of pipradol on higher nervous activity in dogs. Communication I. Effect of piridrol on the strength of inhibition," Trud. Inst. Vyssh. Nervn. Deyat. Ser. Fiziol., 6:300-307.

Mel'nikova, T. A., Zaplatina, O. P., and Kostygov, N. M. (1958). "The action of Anicain and Tropacin on adrenocortical activity," Proceedings of the 7th All-Union Conference of Pharmacologists [in Russian], Khar'kov, pp. 97-98.

Mel'nikova, T. A., Zaplatina, O. P., and Kostygov, N. M. (1959). "The effect of some new cholinolytic agents on adrenocortical function," Probl. Éndokrinol. i Gormonoter., 5(5):14-19.

Meyer, R. I. (1953). "Relative insensitivity of the hypothalamic-pituitary-adrenal system to activation by epinephrine," J. Clin. Endocrinol. Metabol., 13:123-124.

Michaelson, I. A. (1967). "The subcellular distribution of acetylcholine, choline acetyltransferase, and acetyl cholinesterase in nerve tissue," Ann. New York Acad. Sci., 144:387-410.

Mikhailova, N. V. (1955). "Influence of the central nervous system on the adrenocorticotrophic function of the anterior lobe of the pituitary," Prob. Éndokrinol. i Gormonoter., 1(1):59-64.

Mikhel'son M. Ya. (1948). The Action of Narcotics on Cholinsterase [in Russian], Medgiz, Leningrad.

Mikhel'son M. Ya., Zeimal', E. V., Kabanchik, M. I., and Godovikov, N. N. (1961a). "Significance of the onium group in the molecule of anticholinesterase agents for their penetration through the blood-brain barrier (BBB)," Proceedings of the 9th All-Union Pharmacological Conference [in Russian], Sverdlovsk, pp. 172-173.

Mikhel'son, M. Ya., Zeimal', E. V., Fruentov, N. K., and Yakovlev, V. A. (1961b). "Anticholinesterase agents," in: Textbook of Pharmacology [in Russian], Medgiz, Leningrad, pp. 205-237.

Miline, R., Stern, P., and Hulovic, S. (1958). "Sur le variations stressogenes quantitatives de la sérotonine dans le cerveau," Experientia, 14:415-416.

Miyawaki, H., Ui, M., and Kobayashi, B. (1961). "Pituitary adrenocorticotrophic hormone liberation by serotonin," Endocrinol. Japan, 8:148-152.

Moberg, G. P., Scapagnini, U., de Groot, J., and Ganong, W. F. (1971). "Effect of sectioning the fornix on diurnal fluctuation in plasma corticosterone levels in the rat," Neuroendocrinology, 7:11-15.

Mogilevskii, A. Ya. (1960). "Modern views on the role of serotonin in activity of the central nervous system," Uspekhi Sovr. Biol., 50(6):322-336.

Moll, J. (1959). "Localization of brain-stem lesions inhibiting compensatory adrenal hypertrophy following unilateral adrenalectomy," Zschr. Zellforsch., 49:515-524.

Moll. J. (1960). "Localization of hypothalamic lesions inhibiting adrenal weight maintenance in the rat," Acta Endocrinol., 34:19-26.

Moll, J., de Wied, D., and Kranendonk, G. H. (1961). "Observation on the thyroid-adrenal relationship in rats with hypothalamic lesions," Acta Endocrinol., 38:330-340.

Montgomery, R. L., and Berkut, M. K. (1969). "Behavior and adrenal weights of non-stressed rats with selected limbic lobe lesions," Physiol. Behav., 4:745-748.

Morpurgo, C. (1948). "Pharmacological modifications of sympathetic stimulation in the rat," Brit. J. Pharmacol.,34:552-542.

Moussatche, H., and Alvares-Pereiro, H. A. (1957). "Release of adrenocorticotrophine by 5-hydroxytryptamine," Acta Physiol. Latinoamer., 7:71-75.

Müller, J., and Huber, R. (1969). "Effects of sodium deficiency, potassium deficiency and uremia upon the steroidogenic response of rat adrenal tissue to serotonin, potassium and adrenocorticotropin," Endocrinology, 85:43-49.

Muscholl, E. B. (1966). "Indirectly acting sympathomimetic amines," Pharmacol. Rev., 18:551-559.

Nagy, S., Bajusz, G., and Petri, G. (1962). "Response of the adrenal cortex of the dog to surgical injury during autonomic nervous blockade," Acta Physiol. Acad. Sci. Hung., 22:155-162.

Nasmyth, P. A. (1955). "The effect of chlorpromazine on adrenocortical activity in stress," Brit. J. Pharmacol., 10:336-339.

Naumenko, E. V. (1963). "Effect of Piridrol on the pituitary-adrenal system," Izvest. Sibirsk. Otdel. Akad. Nauk SSSR, (12):138-140.

Naumenko, E. V. (1965a). "The role of adrenergic structures in regulation of the function of the pituitary-adrenal system," Proceedings of a Scientific Conference on the Physiology and Pathology of Corticovisceral Relationships and Functional Systems of the Organism [in Russian], Vol. 2, Ivanovo, pp. 96-99.

Naumenko, E. V. (1965b). "The role of cholinergic structures in regulation of the function of the pituitary-adrenal system," Probl. Endocrinol., 11(4):99-104.

Naumenko, E. V (1965c). "The role of adrenergic, cholinergic, and serotoninergic structures in regulation of the activity of the hypothalamo-hypophyseo-adrenal system," Proceedings of an All-Union Conference on the Physiology and Pathology of the Hypothalamus [in Russian], Nauka, Moscow, pp. 195-196.

Naumenko, E. V. (1965d). "Effect of 5-hydroxytryptamine and 5-hydroxytryptophan on the function of the pituitary-adrenal system," Izvest. Sibirsk. Otdel. Akad. Nauk SSSR (3):143-144.

Naumenko, E. V. (1966). "The role of adrenergic, cholinergic, and serotoninergic structures in regulation of the activity of the hypothalamo-hypophyseo-adrenal system," in: The Physiology and Pathology of the Hypothalamus [in Russian], Nauka, Moscow, pp. 150-152.

Naumenko, E. V. (1967a). "Adrenergic and cholinergic structures of the brain in regulation of the pituitary-adrenocortical system," Dokl. Akad. Nauk SSSR, 174(6):1466-1468.

Naumenko, E. V. (1967b). "The role of serotonin in regulation of the activity of the pituitary-adrenal system," Proceedings of an All-Union Scientific Conference on the Biogenic Amines [in Russian], Moscow, pp. 211-212.

Naumenko, E. V. (1967c). "Role of adrenergic and cholinergic structures in the control of the pituitary-adrenal system,' Endocrinology, 80:69-76.

Naumenko, E. V. (1967d). "Adrenergic structures and the control of pituitary-adrenal system," 4th Conf. European Comp. Endocrinol., Abstracts, Addendum, Carlsbad, pp. 13-14.

Naumenko, E. V. (1967e). "Chemoreactive structures and the control of pituitary-adrenal system," Acta Endocrinol., Suppl. 119:42.

Naumenko, E. V. (1968). "Hypothalamic chemoreactive structures and the pituitary-adrenal function. Effect of local injection of norepinephrine, carbachol and serotonin into the brain of guinea pigs with intact brains and after mesencephalic transection," Brain Res., 11:1-10.

Naumenko, E. V. (1969). "Effect of local injection of 5-hydroxytryptamine into rhinencephalic and mesencephalic structures on pituitary-adrenal function in guinea pigs," Neuroendocrinology, 5:81-88.

Naumenko, E. V., and Davydova, E. V. (1965). "The use of Thorn's test in cardiac patients," in: The Pathology of Operations and Anesthesia [in Russian], Nauka, Moscow, pp. 137-140.

Naumenko, E. V., and Il'yuchenok, R. Yu. (1967). "The possibility of a central action of serotonin on the pituitary-adrenocortical system," Byull. Éksperim. Biol. i Med., 64(9):63-66.

Naumenko, E. V., Il'yuchenok, R. Yu., Loskutova, L. V., and Nedbaeva, N. D. (1968). "Serotoninergic structures of the hypothalamus and the function of the pituitary-adrenal system," Probl. Éndokrinol., 14(4):83-86.

Naumenko, E. V., and Popova, N. K. (1965). "Effect of hydrazine derivatives on adrenocortical function," Izvest. Sibirsk. Otdel. Akad. Nauk SSSR, (8):135-138.

Naumenko, E. V., Popova, N. K., and Stargygin, A. G. (1971). "The pituitary-adrenal system of animals in groups and in isolation," Zh. Obshch. Biol., 32:731-739.

Nauta, W. J. H. (1956). "An experimental study of the fornix system in the rat," J. Compar. Neurol., 104:247-270.

Nauta, W. J. H. (1960). "Limbic system and hypothalamus: anatomical aspects," Physiol. Rev., 40:102-104.

Nauta, W. J. H. (1963). "Some connections of the limbic system," in: Mechanisms of the Whole Brain [Russian translation], IL, Moscow, pp. 182-195 (Original version, 1958).

Nelson, D. H., Sandberg, A. A., Palmer, I. G., and Glenn, E. M (1952). "Levels of 17-hydroxycorticosteroids following intravenous infusion of epinephrine into normal men," J. Clin. Endocrinol. Metabol., 12:936-936.

Nesterenko, L. N. (1966). A Study of Correlation between Cholinesterase Activity of the Brain and Blood and Changes in Electrical Activity and Behavior of Animals under the Influence of Tertiary and Quaternary Anticholinesterase Agents, Candidate's Dissertation, Novosibirsk.

Newcomer, W. S. (1962). "Reserpine and adrenocortical function in chicken,," Amer. - J. Physiol., 202:337-339.

Newman, A. E., Redgate, and Farrell, G. (1958). "The effect of diencephalic-mesencephalic lesions on aldosterone and hydrocortisone secretion," Endocrinology, 63:723-736.

Ohler, E. A., and Sevy, R. W. (1953). "Effect of pressor amines and adrenergic blockade on adrenal ascorbic acid concentration in rats," Amer. J. Physiol., 175:285-288.

Ohler, E. A., and Sevy, R. W. (1955). "Blockade of stress-induced adrenal ascorbic-acid depletion," J. Clin. Endocrinol. Metabol., 15:890.

Ohler, E. A., Sevy, R. W., and Weiner, A. (1956). "The effect of chlorpromazine on pituitary-adrenal function," J. Clin. Endocrinol. Metabol., 16:915.

Ohler, E A., and Wakerlin, G. E. (1951). "Amines in experimental hypertensions," J. Lab. Clin. Med., 38:935-936.

Okinaka, S. (1961). "Die Regulation der Hypophysen-Nebennierenfunktion durch das Limbic-System und den Mittelhirnanteil der Formatio-reticularis," Acta Neuroveget. (Vienna), 23:15-20.

Okinaka, S., Ibayashi, H., Motohashi, K., Fujita, T., Yoschida, S., and Ohsaw, N. (1960a). "Effect of electrical stimulation of the limbic system on pituitary-adrenocortical function posterior orbital surface," Endocrinology, 67:319-325.

Okinaka, S., Ibayashi, H., Motohashi, K., Fujita, T., Yoschida, S., Ohsawa, N., and Murakawa, S. (1960b). "Regulation of the pituitary-adrenocortical function through the limbic system and reticular formation," Acta Endocrinol., 35, Suppl. 51:43-44.

Olling, C. J., and de Wied, D. (1956). "Inhibition of the release of corticotrophin from the hypophysis by chlorpromazine," Acta Endocrinol., 22:283-292.

Olling, C. J., and de Wied, D. (1958). "The influence of pacatal on the corticotrophin-releasing activity of the hypophysis," Acta Endocrinol., 28:428-434.

Ondo, J. G., and Kitay, J. I. (1972). "Pituitary-adrenal function in rats with diencephalic islands," Neuroendocrinology, 9:72-82.

Ordy, J. M., Samorajski, T., and Schroeder, D. (1966). "Concurrent changes in hypothalamic and cardiac catecholamine levels after anesthetic, tranquilizers and stress in a subhuman primate," J. Pharmacol. Exptl. Therap., 152:445-457.

Otsuka, K. (1966). "Effects of atropine, eserine and tetramethylammonium on the adrenal 17-hydroxycorticosteroid secretion in anesthetized dogs," Tohoku J. Exptl. Med., 88:165-170.

Paasonen, M. K., and Krayer, O. (1959). "The content of noradrenaline and adrenaline in the rat heart after administration of Rauwolfia alkaloids," Experientia, 15:75-76.

Paasonen, M., McLean, P., and Girman, N. (1957). "5-HT content of structures of limbic system," J. Neurochem., 1:326.

Pace, N., Schaffer, F L., Elmadjian, F., Minard, D., Davis, W., Kilbuck, J. H., Walker, D., Johnston, M. E., Zilinsky, A., Gerard, R. W., Forsham, P. H., and Taylor, J. G. (1956). "Physiological studies on infantryman in combat," California University Publications in Physiology, University of California Press, Berkeley, Vol. 10, pp. 1-88.

Page, I. H. (1954). "Serotonin (5-hydroxytryptamine)," Physiol. Rev., 34:563-588.

Palka, Y., Coyer, D., and Critchlow, V. (1969). "Effect of isolation of medial basal hypothalamus on pituitary-adrenal and pituitary-ovarian function," Neuroendocrinology, 5:333-349.

Pankov, Yu. A. (1961). "Reaction of intact and denervated adrenals of dogs to certain forms of stimulation," Probl. Éndokrinol. i Gormonoter., 7(1):11-15.

Panosyan, G A. (1958). "A photoelectric colorimetric method of determining cholinesterase activity," Izvest. Akad. Nauk Arm. SSR, Biol. S.-Kh. Nauki, 11(6): 21-27.

Parent, A., Saint-Jacques, C., and Poirier, L. J. (1969). "Effect of interrupting on the hypothalamic nervous connections on the norepinephrine and serotonin content of the hypothalamus," Exptl. Neurol., 23:67-75.

Paskov, D. S. (1958). The Pharmacological Characteristics of the Alkaloid Nivaline as an Anticholinesterase Agent, Author's Abstract of Candidate's Dissertation, Leningrad.

Paskov, D. S. (1959). "Nivaline, pharmacological characteristics," Meditsina i Fizkultura (Sofia).

Pastukhov, Yu. F. (1964). An Electroencephalographic Study of the Central Action of Piridrol, Candidate's Dissertation, Novosibirsk.

Paton, W. D. M. (1958). " Central and synaptic transmission in the nervous system," Annual Rev. Physiol., 80:431-470.

Persky, H., Gross, H., Norton, J., and McMurtry, M. (1959). "Effect of hypnotically induced anxiety on the plasma hydrocortisone level of normal subjects," J. Clin. Endocrinol. Metabol., 19:700-710.

Persson, T., and Waldeck, B. (1969). 'The interaction between different metabolic pathways of catecholamines in the brain studied by means of 3H-DOPA," Acta Pharmacol. Toxicol., 27:225-236.

Petropavlovskaya, A. A. (1953). "Reflex hyperglycemia," in: The Pharmacology of New Therapeutic Substances.

Pfeifer, A. K., Vizi, E. S., Satory, E., and Galambos, E. (1963). "The effect of adrenalectomy on the norepinephrine and serotonin content of the brain and on reserpine action in rats," Experientia, 19:482-483.

Pickford, M., and Vogt, M. (1951). "The effect of adrenaline on secretion of cortical hormone in the hypophysectomized dog," J. Physiol. (London), 112:133-141.

Pidevich, I. N. (1971). "Modern views on peripheral serotoninergic structures and serotonin antagonists," in: The Pharmacology of Monoaminergic Processes [in Russian], V. V. Kakusov and N. V. Kaverina (editors), Moscow, pp. 276-313.

Planeles, Kh. Kh., and Popenenkova, Z. A. (1965). Serotonin and Its Role in Infectious Pathology [in Russian], Meditsina, Moscow.

Pletscher, A., and Gey, K. F (1961). "Drug-induced alterations of the metabolism of cerebral monoamines," in: Monoamines et Système Nerveux Central, Geneva.

Pletscher, A., Shore, P., and Brodie, B. (1956). "Serotonin as a mediator of reserpine action in brain," J. Pharmacol. Exptl. Therap., 116:84-89.

Pol, E., Pauž, J., and Sasn, V. (1957. "Controlul functioni suprarenale in urma de ACTH, insulina, adrenalina si pilocarpina administraric," probl. Zootehn. Veterin., 2:49-53.

Polenov, A. L. (1958). "Neurosecretion (secretion of the hypothalamic neurons)," Arkh. Anat. Gistol. i Émbriol., 35(6):114.

Polenov, A. L. (1962). "The functional role of the hypothalamic neurosecretion," Arkh. Anat. Gistol. i Émbriol., 43(9):3-16.

Polenov, A. L. (1968). The Hypothalamic Neurosecretion [in Russian], Nauka, Leningrad.

Poloni, A. (1957). Discussion, in: G. Garattini and V. Ghetti, Psychotropic Drugs, Elsevier, Amsterdam, p. 436.

Popova, N. K. (1971). "Behavior within the group and the pituitary-adrenal system of albino rats," Fiziol. Zh. SSSR, 57:1723-1728.

Popova, N. K., Maslova, L. N., Koryakina, L. A., Bertogaeva, V. D., and Naumenko, E. V. (1972a). "Effect of 5-hydroxytryptophan on the hypothalamo-hypophyseo-adrenal complex after total deafferentation of the mediobasal hypothalamus," Dokl. Akad. Nauk SSSR, 203:726-728.

Popova, N. K., Maslova, L. N., and Naumenko, E. V. (1972b). "Serotonin and the regulation of the pituitary adrenal system after deafferentation of the hypothalamus," Brain Res. (in press).

Popova, N. K., and Naumenko, E. V. (1972). "Dominance relations and the pituitary-adrenal system in rats," Animal Behavior (in press).

Porter, R. W. (1953a). "Neurophysiological aspects of the control of ACTH secretion," in: Third Annual Report on Stress, ACTA, Inc., Montreal, pp. 212-219.

Porter, R. W. (1953b). "Hypothalamic involvement in the pituitary adrenocortical response to stress stimuli," Amer. J Physiol., 172:515-519.

Porter, R. W. (1954). "The central nervous system and stress-induced eosinopenia," Rec. Progr. Horm. Res., 10:1-18.

Porter, I. C. (1963). "Secretion of corticosterone in rats with anterior hypothalamic lesions," Amer. J. Physiol., 204:715-718.

Poskalenko, A. N. (1958). "Effect of difacil on adrenocortical function," in: New Therapeutic Substances in Experimental and Clinical Medicine [in Russian], Izd. Leningrad. S. G. Med. Inst., Leningrad, pp. 29-35.

Poskalenko, A. N. (1960) "Effect of substances acting on adrenocortical function in the region of central cholinergic synapes," Probl. Éndokrinol. i Gormonoter., 6(4):14-17.

Poskalenko, A. N. (1961). "The mechanism of action of difacil (spasmolytin) on adrenocortical function," Byull. Éksperim. Biol. i Med., 51(3):76-79.

Poskalenko, A. N. (1962). "Effect of central cholinolytics on the pituitary-adrenal system," in: The Pharmacology of New Sedatives and Their Clinical Application [in Russian], Medgiz, Leningrad, pp. 129-134.

Poskalenko, A. N. (1965). The Character and Mechanism of Action of Cholinergic Substances on the Hypothalamo-Hypophyseo-Adrenal System, Doctoral Dissertation, Leningrad.

Potter, L. T., and Axelrod, J. (1963). "Studies on the storage of norepinephrine and the effect of drugs," J. Pharmacol. Exptl. Therap., 140:199-206.

Proskurina, N. F., and Yakovleva, A. P. (1952). "The alkaloids of *Galanthus woronowi*. II. Isolation of a new alkaloid," Zh. Obshch. Khim., 22(10):1899-1902.

Proskurina, N. F., and Yakovleva, A. P. (1955). "The alkaloids of *Galanthus woronowi*. III. The structure of galanthamine," Zh. Obshch. Khim., 25(5):1035-1039.

Prozorovskii, V. B. (1960). "Analysis of differences produced by anticholinesterase agents,' Fiziol. Zh. SSSR, 46(5):623-628.

Prozorovskii, V. B. (1961). "Differences in effects of tertiary and quaternary ammonium bases (neostigmine, eserine, methylatropine, and atropine) when administered by different routes," Byull. Éksperim. Biol. i Med., 52(9):73-77.

Put, T. R., and Meduski, I. W. (1962). "The effect of adrenalectomy on the 5-hydroxy-tryptamine metabolism in the rat," Acta Physiol. Pharmacol. Neerl., 11:240-256.

Raile, R. B., Ely, R. S., and Kelley, V. C. (1953). "Studies of 17-hydroxycorticosteroids in children. II. Response of blood levels to the injection of certain hormones," J. Pediat., 42:46-53.

Ramey, E. R., and Goldstein, M. S. (1957). "The adrenal cortex and sympathetic nervous system," Physiol. Rev., 37:155-195.

Rasin, M. S., Braude, I. Ya., and Simon, I. B. (1970). "Character of adrenoreception in the hypothalamo-hypophyseo-adrenal system," Probl. Éndokrinol., 16(5):42-45.

Rausch-Strooman, J. G., and Witthöft, R. (1956). "Vergleichende Untersuchungen über des Einfluss von Isoniazid und ACTH auf die Steroidausscheidung der Nebennierenrinde," Klin. Wschr., 34:140-142.

Recant, L., Hume, D. M., Forshman, P. H., and Thorn, G. W. (1950). "Studies of the effect of epinephrine on the pituitary-adrenocortical system," J. Clin. Endocrinol. Metabol., 19:187-229.

Redding, T. W., Bowers, C. Y., and Schally, A. V. (1966). "An *in vitro* assay for thyrotropin releasing factor," Endocrinology, 79:229-236.

Redgate, E. S. (1970). "ACTH release evoked by electrical stimulation of brain stem and limbic system sites in the cat: the absence of ACTH release upon infundibular area stimulation," Endocrinology, 86:806-823.

Réithelyi, M., and Halász, B. (1970). "Origin of the nerve endings in the surface zone of the median eminence of the rat hypothalamus," Exp. Brain Res., 11:145-158.

Resnick, R. H., Smith, G. T., and Gray, S. I. (1961). "Endocrine influences on tissue serotonin content of the rat," Amer. J. Physiol., 201:571-573.

Revzin, A. M., and Maickell, R. (1966). "Failure of fornix section to block ACTH hypersecretion elicited by reserpine or cold exposure," Life Sci., 5:253-257.

Rezabek, K., and Votava, Z. (1956). "The action of some antihistaminics and neuroplegics on the release of ACTH in surgical stress in rats," Arch. Internat. Pharmacodyn. Therap., 108:279-288.

Richter, D., and Crossland, J. (1949). "Variation in acetylcholine content of the brain with physiological state," Amer. J. Physiol., 159:247-255.

Riege, W. H., and Morimoto, H. (1970). "Effect of chronic stress and differential environments upon brain weights and biogenic amine levels in rat," J. Comp. Physiol. Psychol., 71:396-404.

Rikkl', A. V. (1934). "The neuro-humoral transmission of excitation," Arkh. Biol. Nauk. Ser. A, 35(3):349-356.

Riley, H. A. (1960). An Atlas of the Basal Ganglia, Brain Stem, and Spinal Cord, Hafner, New York.

Rinne, U. K., and Sonninen, V. (1968). "The occurrence of dopamine and noradrenaline in the tubero-hypophyseal system," Experientia, 24:177-178.

Robertis, E. de (1961). "Morphological bases of synaptic processes and neurosecretion," in: Regional Neurochemistry, S. S. Kety and J. Elkes (editors), Pergamon Press, New York, pp. 248-257.

Robertis, E. de, and Fiszer, S. (1968). "Distribution of acetylcholinesterase, proteolipids, and cholinergic receptor in cat brain," J. Pharm. Pharmacol., 20:146-148.

Robertis, E. de, Iraldi, A. P, Lores Arnaiz, G. R., and Zieher, L. M. (1965). "Synaptic vesicles from the rat hypothalamus. Isolation and norepinephrine content," Life Sci., 4:193-201.

Roberts, E., and Eidelberg, E. (1960). "Metabolic and neurophysiological roles of γ-aminobutyric acid," Internat. Rev. Neurobiol., 2:280-327.

Roberts, S., and Keller, M. R. (1955). "Influence of epinephrine and cortisone on the metabolism of the hypophysis and hypothalamus of the rat," Endocrinology, 57:64-69.

Romero, E., and Solohaga, A. (1961). "Estudio de los 17-cetosteroides por la accion de las hydrazidas," Rev. Clin. Espan., 82:232-236.

Ronzoni, E., and Reichlin, S. (1950). 'Adrenergic agents and the adrenocorticotrophic activity of the anterior pituitary," Amer. J. Physiol., 160:490-498.
Rosecrans, J. A. (1969). "Brain amine changes in stressed and normal rats pretreated with various drugs," Arch. Int. Pharmacodyn., 180:460-470.
Rosecrans, J. A. (1970). "Effects of acute stress on forebrain 5-hydroxytryptamine metabolism and pituitary adrenal function," Europ. J. Pharmacol., 9:170-174.
Rosendorff, C., and Cranston, W. I. (1971). "Effects of intrahypothalamic and intraventricular norepinephrine and 5-hydroxytryptamine on hypothalamic blood flow in the conscious rabbit," Circulation Res., 28:492-502.
Rosenfeld, G. (1955). "Stimulative effect of acetylcholine on the adrenocortical function of isolated perfused calf adrenals," Amer. J. Physiol., 183:272-278.
Rossum, J. M. van (1963). "Significance of dopamine in psychomotor stimulant action, in: Biochemical and Neurophysiological Correlation of Centrally Acting Drugs, E. Trabbucchi, R. Paoletti, and N. Canal (editors), Pergamon Press, Oxford, pp. 115-126.
Rubin, R. M., Mandell, A. J., and Crandell, P. H. (1966). "Corticosteroid responses to limbic stimulation in man: localisation of stimulus sites," Science, 153:767-768.
Ryzhenkov, V. E. (1958). "Reflexes from chemoreceptors of the carotid bodies to the adrenal cortex," Abstracts of Proceedings of the 7th All-Union Conference of Pharmacologists [in Russian], Khar'kov, p. 127.
Ryzhenkov, V. E (1959a). "Reflexes from chemoreceptors on adrenocortical activity," Probl. Endokrinol. i Gormonoter., 5(1):39-43.
Ryzhenkov, V. E. (1959b). "Role of reflexes from the carotid chemoreceptors in the action of nicotine, corconium, and sodium sulfide on the secretion of 17-hydroxycorticosteroids by the adrenals in dogs," Probl. Éndokrinol. i Gormonoter, 5(6): 19-23.
Saba, G. C., Carnicelli, A., Saba, P., and Marescotti, V. (1963). "Diurnal rhythm in the adrenal cortical secretion and in the rate of metabolism of corticosterone in the rat,'' Acta Endocrinol., 44:413-415.
Saba, P., Saba G. C., and Bergamini, E. (1962a). "Effect of hypothalamic lesions on the adrenal cortex in rat. Relationships between the plasma corticosterone levels, the mean nuclear diameters of the cells of the zona fasciculata and the nuclear mass of the adrenal cortical cells," Folia Endocrinol. (Rome), 15:681-684.
Saba, G. C., Saba, P., and Carnicelli, A. (1962b). "Effects of oxytocin, vasopressin and adrenaline on the adrenal cortical function in rats with lesions in the eminentia mediana," Folia Endocrinol. (Rome), 15:685-687.
Saba, G. C., Saba, P., and Marescotti, V. (1962c). "Effects of hypothalamic lesions on corticosterone production," Folia Endocrinol. (Rome), 15:676-680.
Saffran, M., Schally, A. V., and Benfrey, B. G. (1955). "Stimulation of the release of corticotropin from the adenohypophysis by a neurohypophysial factor," Endocrinology, 57:439-444.
Sager, O. (1962). The Diencephalon [Russian translation], Rumanian Academy of Sciences Press (original version, 1960).
Salmoiraghi, G. C. (1966). "Central adrenergic synapses," Pharmacol. Rev., 18:717-726.
Salmoiraghi, G. C., Costa, E., and Bloom, F. E. (1965). "Pharmacology of central synapses," Annual Rev. Pharmacol., 5:213-234.

Salmoiraghi, G. C., and Stefans, C. N. (1965). "Patterns of central neuron responses to suspected transmitters," Arch. Ital. Biol., 103:705-724.

Samuels, A. I. (1951). "Primary and secondary leukocyte changes following the intra-muscular injection of epinephrine hydrochloride," J. Clin. Invest., 30:941-947.

Sandberg, A. A., Nelson, D. H., Palmer, I. G., Samuels, L. T., and Tyler, F. H. (1953). "The effect of epinephrine on the metabolism of 17-hydroxycorticosteroids in the human," J. Clin. Endocrinol. Metabol., 13:629-647.

Sanzari, N. P., Possanza, G. J., and Troop, R. C. (1965). "Lack of correlation between body weight and cortisol secretion, and between adrenal weight and cortisol secretion in the dog," Federat. Proc., 24:574.

Sapeika, N. (1959). "The effect of chlorpromazine, iproniazid, and chloroquine on adrenal ascorbic acid in the rat,' Arch. Internat. Pharmacodyn. Therap., 122: 196-200.

Sawyer, C. H. (1952). "Specific and nonspecific blocking effects of dibenamine congeners on the stress response in rats," J. Clin. Endocrinol. and Metabol., 12:959-960.

Sawyer, C. H , and Parkerson, G. R. (1953). "Mechanisms of partial blockade of the stress response in rats by dibenamine analogues," Endocrinology, 52:346-356.

Sayers, G. (1960). "Hypothalamus and adenohypophysis: with special reference to corticotrophin release," Acta Endocrinol., Suppl. 50:25-31.

Sayers, G., Redgate, E. S., and Royce, P.C. (1958). "Hypothalamus, adenohypophysis and adrenal cortex," Annual Rev. Physiol., 20:243-274.

Sayers, G., and Sayers, M. A. (1947). "Regulation of pituitary adrenocorticotrophic activity during the response of the rat to acute stress," Endocrinology, 40:265-273.

Sayers, G., and Sayers, M. A. (1948). "The pituitary-adrenal system," Rec. Progr. Horm. Res., 2:81-106.

Scaltrine, G. C. (1956). "Il quadro emocitario periferico e la glicemia dopo 5-idrossitriptamina," Haematologica, 41:681-687.

Scapagnini, U., Moberg, G. P., Van Loon, G. R., De Groot, J., and Ganong, W. F. (1971). "Relation of brain 5-hydroxytryptamine content to the diurnal variation in plasma corticosterone in the rat," Neuroendocrinology, 7:90-96.

Scapagnini, U., Van Loon, G. R., Moberg, G. P., and Ganong, W. F. (1970). "Effect of α-methyl-p-tyrosine on the circadian variation of plasma corticosterone in rats," Europ. J. Pharmacol., 11:266-268.

Schäfer, E. L. (1961). "Die Zuverlässigkeit einiger Funktionszüfugen der Nebenniere," Dtsch. Med. Wochenschr., 86:2257-2265.

Schallek, W., and Zabransky, F. (1966). "Effects of psychotropic drugs on pressor responses to central and peripheral stimulation in cat," Arch. Internat. Pharmacodyn. Therap., 161:126-131.

Schally, A. V., Anderson, R. N., Lipscomb, H. S., Long, I. M., and Guillemin, R. (1960). "Evidence for the existence of two corticotrophin-releasing factors α-CRF and β-CRF," Nature, 188:1192.

Schally, A. V., and Guillemin, R. (1960). "Studies on corticotrophin-releasing factor: ion exchange chromatography of pituitary preparations," Texas Rep. Biol. Med., 18:133-146.

Schally, A. V., Lipscomb, H. S., and Guillemin, R. (1962a). "Isolation and amino acid sequence of α_2-corticotrophin-releasing factor (α_2-CRF) from dog pituitary glands," Endocrinology, 71:164-173.

Schally, A. V., Lipscomb, H. S., Long, I. M., Dear, W. E., and Guillemin, R. (1962b). "Chromatography and hormonal activities of dog hypothalamus," Endocrinology, 70:478-480.

Schapiro, Sh., and Aragonas, E. (1963). "Ascorbic acid and corticosterone concentration of innervated and denervated adrenal glands," Endocrinology, 72:836-837.

Scharrer, E. (1965). "The final common path in neuroendocrine integration," Arch. Anat. Microsc. Morphol. Exptl., 54:359-370.

Scharrer, E., and Scharrer, B. (1940). "Secretory cells within the hypothalamus," Res. Publ. Assoc. Nerv. Ment. Dis., 20:170-194.

Scharrer, E., and Scharrer, B. (1954). "Die Neurosekretion," in: Handbuch der mikroskopischen Anatomie des Menschen, Springer, Berlin, Vol. 6, pp. 953-1066.

Scharrer, E., and Scharrer, B. (1963). Neuroendocrinology, Columbia University Press, New York.

Schmid, R., Gonzalo, L., Blobel, R., Muschke, E., and Tonutti, E. (1957). "La partecipazione dell'ipotalamo nella mobilizazione dell'ACTH," Folia Endocrinol. (Rome), 10:3-9.

Schmid, E., Scheiffarth, F., Zicha, L., and Schwemmle, C. (1960a). "Wechselbeziehungen zwischen 5-Hydroxytryptamin und dem Hypophysen-Nebennierenrinden-System," Acta Endocrinol., 51:843.

Schmid, E., Scheiffarth, F., Zicha, L., and Siede, H. (1960b). "Untersuchengen über die Rolle des 5-Hydroxytryptamin bei allergischen und rheumatischen Erkranken sowie über die Beeinflussbarkeit des Blutserotoninspiegels durch ACTH und Glucocorticoide," Z. Ges. Exptl. Med., 133:1-8.

Schneider, H. P. G., and McCann, S. M. (1970). "Mono- and indolamines and control of LH-secretion," Endocrinology, 86:1127-1133.

Schweizer, M. (1953). "Alterations in the white blood cell picture in normal, hypophysectomized and splenectomized guinea pigs during mild stress and following epinephrine injection," Endocrinology, 53:293-300.

Schweizer, A., Stedman, E., and Wright, S. (1939). "Central action of anticholinesterases," J. Physiol. (London), 96:302-336.

Schwemmle, C., Schmid, F., Scheiffarth, F , Zicha, L., and Schichor, E. (1961). "Zur Frage der gegenseitigen Beziehung zwischen adrenocorticotropem Hormon bzw. Glucocorticoiden und dem Serotonin-Stoffwechsel," Arzneimittel-Forsch., 11:616-619.

Seggie, J., and Brown, G M. (1971). "Septal lesions and resting adrenal function. A possible explanation of conflicting findings," Neuroendocrinology, 8:367-374.

Selye, H. (1937). "Studies on adaptation," Endocrinology, 21:169-188.

Selye, H. (1960). "The Story of the Adaptation Syndrome [Russian translation], Medgiz, Moscow (Original version, 1952).

Setekleiv, I., Skaug, O. E., and Kaada, B. R. (1961). "Increase of plasma 17-hydroxycorticosteroids by cerebral cortical and amygdaloid stimulation in the cat," J. Endocrinol., 22:119-127.

Sevy, R. W., Ohler, E. A., and Weiner, A. (1957). "Effect of chlorpromazine on stress-induced adrenal ascorbic acid depletion," Endocrinology, 61:45-51.

Shalyapina, V. G. (1967a). "The mechanism of the reverse action of corticosteroids and the role of the brain catecholamines in it," Proceedings of an All-Union Scientific Conference on the Biogenic Amines, Moscow, pp. 321-322.

Shalyapina, V. G. (1967b). "The role of the brain noradrenalin in the regulation of the pituitary-adrenal system," in: Proceedings of an All-Union Scientific Conference on the Biogenic Amines [in Russian], Moscow, p. 322.

Shalyapina, V. G., and Rakitskaya, V. V. (1971). "Central monoaminergic structures and their role in the regulation of stress reactions (a survey of the literature)," Probl. Éndokrinol., 17(2):107-114.

Share, N. N., and Melville, K. G. (1963). "Centrally mediated sympathetic cardiovascular responses induced by intraventricular norepinephrine," J. Pharmacol. Exptl. Therap., 141:15-21.

Shchedrina, R. N. (1971). "Relations between the catecholamines of the hypothalamus and the adrenocorticotrophic function of the pituitary after adrenalectomy," Probl. Éndokrinol., 17(1):58-64.

Sheard, M. H., and Aghajanian, G. K. (1967). "Neural release of brain serotonin and body temperature," Nature, 216:495-496.

Shore, P. A. (1958). "Possible mechanism of antidepressant action of marsilid," J. Clin. Exptl. Psychopathol., 19, Suppl. 1:56-60.

Shreiberg, G. L. (1966). "The neuro-humoral mechanisms of regulation of the function of the hypothalamus–pituitary–adrenal cortex system," in: The Physiology and Pathology of the Hypothalamus [in Russian], Nauka, Moscow, pp. 30-38.

Shreiberg, G. L., and Dunaeva, L. P. (1965). "The mediator nature of the nervous structures of the hypothalamus participating in regulation of the function of the hypothalamus–pituitary–adrenal cortex system," in: Physiology and Pathology of the Endocrine System. Proceedings of a Congress of Endocrinologists of the Ukrainian SSR [in Russian], Khar'kov, pp. 537-538.

Shreiberg, G. L., and Matlina, E. Sh. (1967). "Effect of injection of serotonin into the rat hypothalamus on the catecholamine concentration in the adrenals," Proceedings of an All-Union Scientific Conference on the Biogenic Amines [in Russian], Moscow, pp. 323-324.

Siderius, P., and Gaarenstroòm, J. H. (1952). "The effect of dihydroergotamine on the blood count and adrenal cortical response to adrenaline," Acta Endocrinol., 9:109-117.

Similä, S. J., and Laine, M. J. (1962). "Über die Wirkung von Isoniazid und 2-Athylisothionicotinamid auf das hypophysär-adrenocorticale System der Ratte," Beitr. Klin. Tuberkul., 126:234-238.

Slusher, M. A. (1958). "Dissociation of adrenal ascorbic acid and corticosterone responses to stress in rats with hypothalamic lesions," Endocrinology, 63:412-419.

Slusher, M. A. (1970). "Effect of brain stem lesions on stress-induced corticosteroid release in female rats," Endocrinology, 67:347-353.

Slusher, M. A. (1964). "Effects of chronic hypothalamic lesions on diurnal and stress corticosteroid levels," Amer. J. Physiol., 206:1161-1164.

Slusher, M. A., and Critchlow, V. (1959). "Effect of midbrain lesions on ovulation and adrenal response to stress in female rats," Proc. Soc. Exptl. Biol. (New York), 101:497-499.

Slusher, M. A., and Hyde, I. E., (1961a). "Inhibition of adrenal corticosteroid release by brain stem stimulation in cats," Endocrinology, 68:773-782.

Slusher, M. A., and Hyde, I. E. (1961b). "Effect of limbic stimulation on release of corticosteroids into the adrenal venous effluent of the cat," Endocrinology, 69:1080-1084.

Slusher, M. A., and Hyde, J. E. (1966). "Effect of diencephalic and midbrain stimulation on ACTH levels in unrestrained cats," Amer. J. Physiol., 210:103-108.

Sly, R. M., and Heimlich, E. M. (1966). "A review: sympathetic and sympathomimetic action," Ann. Allergy, 24:259-272.

Smelik, P. G. (1967). "ACTH secretion after depletion of hypothalamic monoamines by reserpine implants," Neuroendocrinology, 2:247-254.

Smelik, P., and de Wied, D. (1958). "Corticotrophin-releasing action of adrenaline, serotonine and pitressin," Experientia, 14:17.

Smirnov, G. D. (1957). Electrical Phenomena in the Central Nervous System and Their Changes due to the Effects of Certain Pharmacochemical Agents on Tissue Metabolism, Author's Abstract of Doctoral Dissertation, Moscow.

Smith, G. P., and Root, A. W. (1971). "Dissociation of changes in growth hormone and adrenocortical hormone levels during brain stimulation of monkeys," Neuroendocrinology, 8:235-244.

Smookler, H. H., and Buckley, J. P. (1969). "Relationships between brain catecholamine synthesis, pituitary adrenal function and the production of hypertension during prolonged exposure to environmental stress," Int. J. Neuropharmacol., 8:33-41.

Snyder, J., and D'angleo, S. A. (1963). "Hypothalamic stimulation and ascorbic acid content of endocrine glands of the albino rat," Proc. Soc. Exptl. Biol. Med., 112:1-4.

Sourkes, T. L. (1964). "Actions of dopa and dopamine in relation to function of the central nervous system," in: Biochemical and Neurophysiological Correlation of Centrally Acting Drugs, 2nd International Pharmacological Meeting, E. Trabucchi, R. Paoletti, and N. Canal (editors), Pergamon Press, Oxford, Vol. 2, pp. 35-50.

Spector, S., Shore, P. A., and Brodie, B. B. (1960). "Biochemical and pharmacological effects of monoamineoxidase inhibitors: iproniazid, 1-phenyl-2-hydrazinopropane (JB 516) and 1-phenyl-3-hydrazinobutane (JB 835)," J. Pharmacol. Exptl. Therap., 128:15-21.

Speirs, R. S. (1953). "Eosinopenic activity of epinephrine in adrenalectomized mice," Amer. J. Physiol., 172:520-526.

Speirs, R. S., and Meyer, R. K. (1949). "The effects of stress, adrenal and adrenocorticotrophic hormones on the circulating eosinophils of mice," Endocrinology, 45:403-429.

Steiner, F. A., and Hedinger, Chr. (1956). "Thrombozytose und Eosinopenie bei Ratten nach einmaliger 5-Hydroxytryptamininjektion," Experientia, 12:109-110.

Steiner, F. A., Siebenmann, R. E., Sandri, C., and Hedinger, Chr., (1957). "Thrombozytose und Eosinophilie bei adrenalektomierten Ratten nach einmaliger 5-Hydroxytryptamininjektion," Experientia, 13:500-502.

Stone, W. E. (1957). "The role of acetylcholine in brain metabolism and function," Amer. J. Phys. Med., 36:222-255.

Stoner, H. B., and Elson, P. M. (1971). "The effect of injury on monoamine concentrations in the rat hypothalamus," J. Neurochem., 81:1837-1846.

Suzuki, T., Hirai, K., Yoshio, H., Kurouji, K.-I., and Hirose, T. (1964). "Effect of eserine and atropine on adrenocortical hormone secretion in unanesthetized dogs," J. Endocrinol., 31:81-82.

Suzuki, T., Romanoff, E. B., Koella, W. P., and Levy, C. I. (1960). "Effect of diencephalic stimulation on 17-hydroxycorticosteroid secretion in unanesthetized dog," Amer. J. Physiol., 198:1312-1314.

Svirchevskaya, V. V. (1958). "Changes in the blood composition during treatment of patients with chronic polyarthritis by means of adrenocorticotrophic hormone and cortisone," Probl. Éndokrinol. i Gormonoter., 4(4):65-71.

Swamy, V. C., Tye, A., Lapidus, J. B., and Patil, P. N. (1967). "Potentiation of norepinephrine responses by isomers of amphetamine," Life Sci., 6:903-911.

Symington, T. (1951). "The reaction of the adrenal cortex to adrenaline stress and identification of endogenous liberated sugar-active corticoid by a histological-bio-assay technique," Brit. J. Exptl. Pathol., 32:58-69.

Szentagothai, J., Flerko, B., Més, B., and Halász, B. (1965). Hypothalamic Regulation of the Anterior Pituitary, Hungarian Academy of Sciences Press, Budapest.

Taylor, A. N., and Farrell, G. (1962). "Effect of brain stem lesions on aldosterone and cortisol secretion," Endocrinology, 70:556.

Telford, I. M., and West, G. B. (1960). "The effects of corticosteroids and related compounds on the histamine and 5-hydroxytryptamine content of rat tissues," Brit. J. Pharmacol., 15:532-539.

Tepperman, J., and Bogardus, J. S. (1948). "Attempts at pharmacologic blockade of the secretion of adrenocorticotrophin," Endocrinology, 43:448-450.

Tepperman, J., and Tepperman, H. M. (1950), "Metabolic function of the endocrine glands," Annual Rev. Physiol. 12:503-536.

Thierry, A. M., Fekete, M., and Glowinski, G. (1968). "Effects of stress on the metabolism of noradrenaline, dopamine and serotonin (5-HT) in the central nervous system of the rat. (II). Modifications of serotonin metabolism," Europ. J. Pharmacol., 4:384-389.

Thorn, G. W., and Forsham, P. H. (1949). "Metabolic changes in man following adrenal and pituitary hormone administration," Rec. Progr. Horm. Res., 4:229-288.

Tigyi, A., Puppi, A., and Lissak, K. (1959). "The role of adrenaline and noradrenaline in adaptive reactions," Acta Physiol. Acad. Sci. Hung., 16:41-45.

Titus, E., and Udenfriend, S. (1954). "Metabolism of 5-hydroxytryptamine (serotonin), Federat. Proc., 13:411.

Toda, N., Matsuda, Y., and Shimamoto, K. (1969). "Cardiovascular effects of sympathomimetic amines injected into the cerebral ventricles of rabbits," Int. J. Neuropharmacol., 8:451-461.

Toh, C. C. (1960). "Effect of temperature on the 5-hydroxytryptamine (serotonin) content of tissue," J. Physiol. (London), 151:410-415.

Tonkikh, A. V. (1968). The Hypothalamo-Hypophyseal Region and Regulation of the Physiological Functions of the Organism [in Russian], Nauka, Leningrad.

Tonutti, E. (1961). "Experimente zur Lokalisation der hypothalamischen Steurung der ACTH-Sekretion," Acta Neuroveget. (Vienna), 23:35-49.

Towne, I. C., and Sherman, I. O. (1960). "Failure of acute bilateral adrenalectomy to influence brain serotonin levels in the rat," Proc. Soc. Exptl. Biol. (New York), 103:721-722.

Tronchetti, F., Marescotti, V., Saba, G. C., Carnicelli, A., and Saba, P. (1965). "Investigation on control of adrenocortical activity by the nervous system," Europ. Rev. Endocrinol., 1:281-290.

Tronchetti, F., Marescotti, V., Saba, G. C., and Saba, P. (1964). "Optic pathway and hypothalamic structures: their role in the control of adrenal function," Proc. 2nd Internat. Congr. Endocrinol., London, pp. 538-543.

Trubitsyna, T. K. (1960). "The vasoconstrictor naphazoline," Med. Prom. SSSR, (12):49-51.

Tsoi, S. A. (1964). "The functional and morphological state of the adrenal cortex under the influence of some central neurotropic agents," Probl. Éndokrinol. i Gormonoter., 10(6):66-70.

Udenfriend, S. (1958). "Metabolism of 5-hydroxytryptamine," in: 5-Hydroxytryptamines, Lewis (editor), Pergamon Press, Oxford, pp. 43-59.

Udenfriend, S., Weissbach, H., and Bogdanski, D. F. (1957a). "Biochemical findings relating to the action of serotonin," Ann. New York Acad. Sci., 66:602-608.

Udenfriend, S., Weissbach, H., and Bogdanski, D. F. (1957b). "Increase in tissue serotonin following administration of its precursor 5-hydroxytryptamine," J. Biol. Chem., 224:803-810.

Usher, D. R., Kasper, P., and Birmingham, M. K. (1967). "Comparison of pituitary-adrenal function in rats lesioned in different areas of the limbic system and hypothalamus," Neuroendocrinology, 2:157-174.

Uscher, D. R., and Lamble, R. W. (1969). "ACTH synthesis and release in septal-lesioned rats exposed to air shuttle-avoidance," Physiol. Behav., 4:923-927.

Utevskii, A. M., and Kulinskii, V. I. (1967). "Catecholamines in research by Soviet scientists," Probl. Éndokrinol., 13(5):84-97.

Valzelli, L., and Garattini, S. (1968). "Biogenic amines in discrete brain areas after treatment with monoamineoxidase inhibitors," J. Neurochem., 15:259-261.

Van Loon, G. R., Hilger, L., Cohen, R., and Ganong, W. F. (1969). "Evidence for a hypothalamic adrenergic system that inhibits ACTH secretion in the dog," Fed. Proc., 28:438.

Van Loon, G. R., Hilger, L., King, A. B., Boryczka, A. T., and Ganong, W. F. (1971a). "Inhibitory effect of L-dihydroxyphenylalanine on the adrenal venous 17-hydroxycorticosteroid response to surgical stress in dogs," Endocrinology, 88: 1404-1414.

Van Loon, G. R., Scapagnini, U., Cohen, R., and Ganong, W. F. (1971b). "Effect of intraventricular administration of adrenergic drugs on the adrenal venous 17-hydroxycorticosteroid response to surgical stress in the dog," Neuroendocrinology, 8:257-272.

Van Loon, G. R., Scapagnini, U., Moberg, G. P., and Ganong, W. F. (1971c). "Evidence for central adrenergic neural inhibition of ACTH secretion in the rat," Endocrinology, 89:1464-1469.

Varagic, V., Krstic, M., and Beleslin, B. (1966). "The effect of acute electrolytic lesions in the hypothalamus on the hypertensive response to physostigmine in the rat," Arch. Internat. Pharmacodyn. Therap., 162:162-169.

Vekshina, N L. (1967). "Changes in the blood serotonin concentration in rabbits during prolonged electrical stimulation of the anterior hypothalamic zones," Byull. Éksperim. Biol. i Med., 63(5):7-10.

Vening, E. H. (1965). "Adenohypophysis and adrenal cortex," Annual Rev. Physiol., 27:107-132.

Verdesca, A. S., Westermann, C. D., Crampton, R. S., Black, W. C., Nedeljkovic, R. I., and Hilton, J. I. (1961). "Direct adrenocortical stimulatory effect of serotonin," Amer. J. Physiol., 201:1065-1067.

Vernikos-Danellis, J. (1964). "Neuroendocrine factors affecting the synthesis and release of ACTH," in: Proc. 2nd Internat. Congr. Endocrinol., Amsterdam, pp. 549-555.

Vernikos-Danellis, J., and Marks, B. H. (1962). "Epinephrine-induced release of ACTH in normal human subjects: a test of pituitary function," Endocrinology, 70:525-531.

Vinogradskii, A. V. (1965). "Effect of some neurotropic agents (phenobarbital and amphetamine) on the functional state of the adrenal cortex," Trudy 1-go Mosk. Med. Inst., 43:103-112.

Vogt, M. (1944). "Observations on some conditions affecting the rate of hormone output by the surrenal cortex," J. Physiol. (London), 103:317-332.

Vogt, M. (1947). "Some aspects of the physiology of the secretion of the adrenal cortex with a bearing on clinical medicine," Exptl. Med. Surg., 5:279-284.

Vogt, M. (1951). "Cortical secretion of the isolated perfused adrenal," J. Physiol. (London), 113:129-156.

Vogt, M. (1955). "Medullary-cortical relationships in the adrenal," Ciba Found. Coll. Endocrinol. (London), 8:241-248.

Vogt, M. (1957). "Distribution of adrenaline and noradrenaline in the cerebral nervous system and its modification by drug," in: Metabolism of the Nervous System, pp. 553-565.

Vogt, M. (1960). "The control of the secretion of corticosteroids," in: Biochem. Soc. Symposia, Cambridge, 18:85-95.

Vogt, M. (1964). "The effect of chlorpromazine and reserpine on the release of ACTH," Proc. 2nd Internat. Congr. Endocrinol. Excerpta Medica Internat. Congress, Series No. 83, pp. 635-639.

Voitkevich, A. A. (1960). "Neurosecretion and the endocrine glands," Current Problems in Endocrinology, No. 1, pp. 48-91.

Voitkevich, A. A. (1961). "The neurosecretory component of the adaptation reaction," Patol. Fiziol. i Éksperim. Terapiya, 5(6):73-80.

Voitkevich, A. A. (1964). "The hypothalamo-hypophyseal neurosecretory system and experimental ways of its investigation," in: The Neurosecretory Elements and Their Role in the Organism [in Russian], Nauka, Moscow–Leningrad, pp. 72-74.

Votava, Z. (1967). "Pharmacology of the central cholinergic synapses," Ann. Rev. Pharmacol., 7:223-240.

Waldron, H. A. (1969). "Cholinergic fibres in the spinal cord of the guinea pig," Brain Res., 12:250-252.

Waund, D., Kottegoda, S. R., and Krayer, O. (1958). "Threshold dose and time course of norepinephrine depletion of the mammalian heart by reserpine," J. Pharmacol. Exptl. Therap., 12:340.

Way, E. L., and Sutherland, V. C. (1963). "Pharmacologically active brain substances and their relation to endocrine effects," Anesthesiology, 24:543-562.

Way, E. L., Taylor, I., Old, L. J., and George, R. (1962). "Corticotrophic studies with aspirin," J. Endocrinol., 24:7-16.

Weight, F. F., and Salmoiraghi, G. C. (1966). "Adrenergic responses of Renshaw cells," J. Pharmacol. Exptl. Therap., 154:391-397.

Weil-Malherbe, H. (1960). "The passage of catecholamines through the blood–brain barrier," in: Ciba Foundation Symposium Jointly with the Committee for Symposia on Drug Action on Adrenergic Mechanisms, Vanl (editor), London, pp. 421-423.

Weil-Malherbe, H., Axelrod, J., and Tomchick, R. (1959). Blood–brain barrier for adrenaline," Science, 129:1226-1227.

Weiner, N. (1960). "The distribution of monoamine oxidase and succinic oxidase in brain," J. Neurochem., 6:79-86.

Weiner, N. (1964). "The catecholamines: biosynthesis, storage and release, metabolism and metabolic effects," in: The Hormones, Academic Press, New York, Vol. 4, pp. 403-479.

Weiner, R. I., Shryne, J. E., Gorski, R. A., and Sawyer, C. H. (1972). "Changes in the catecholamine content of the rat hypothalamus following deafferentation," Endocrinology, 90:867-873.

Weissbach, H., Lovenberg, W., Redfield, B. G., and Udenfriend, S. (1961). "*In vivo* metabolism of serotonin and tryptamine: effect of monoamine oxidase inhibition," J. Pharmacol. Exptl. Therap., 131:26-30.

Welch, A. S., and Welch, B. L. (1968). "Reduction of norepinephrine in the lower brain stem by psychological stimulus," Proc. Nat. Acad. Sci. (Washington), 60: 478-481.

Welch, B. L. (1965). "Psychological response to the mean level of environmental stimulation; a theory of environmental integration," in: Symposium on Medical Aspects of Stress in the Military Climate, Apr. 22-24, 1964, U. S. Government Printing Office Publ., pp. 39-99.

Wesemann, W., Henkel, R., and Marx, R. (1971). "Receptors of neurotransmitters. V. Sialic acid distribution and characterization of the 5-hydroxytryptamine receptor in synaptic structures," Biochem. Pharmacol., 20:1961-1966.

Westermann, E. O. (1965). "Cumulative effects of reserpine on the pituitary-adrenocortical and sympathetic nervous system," in: Drugs and Enzymes, Pergamon Press, Oxford, pp. 381-392.

Westermann, E. O., Maickel, R. P., and Brodie, B. B. (1962). "On the mechanism of pituitary-adrenal stimulation by reserpine," J. Pharmacol. Exptl. Therap., 138: 208-217.

Whittaker, V. P. (1961). "The binding of neurohormones by subcellular particles of brain tissue," in: Regional Neurochemistry, S. S. Kety and J. Elkes (editors), Pergamon Press, New York, pp. 259-262.

Whittaker, V. P. (1966). "Catecholamine storage particles in the central nervous system," Pharmacol. Rev., 18:401-412.

Wied, D. de (1961). "Effect of hypothalamic lesions and hypophysectomy on corticoid production *in vivo* and on adrenal weight in rats," Acta Endocrinol., 37:279-287.

Wied, D. de (1967). "Chlorpromazine and endocrine function," Pharmacol. Rev., 19:251-288.

Wise, B. L., Van Brunt, E. E., and Ganong, W. F. (1963). "Effect of removal of various parts of the brain on ACTH secretion in dogs," Proc. Soc. Exptl. Biol. (New York), 112:792-795.

Wise, B. L., Van Brunt, E. E., and Ganong, W. F. (1964). "Effect of brain removal in dogs previously subjected to pituitary stalk section," Proc. Soc. Exptl. Biol. (New York), 116:306-310.

Wolfson, W. Q. (1953). "Inadequacy of epinephrine as an activator of the pituitary-adrenal system," J. Clin. Endocrinol. Metabol., 13:125.

Wolley, D. W., and Gommi, B. W. (1966). "Serotonin receptors. VI. Methods for the direct measurement of isolated receptors," Arch. Internat. Pharmacodyn. Therap., 159:8-17.

Yamada, T., and Greer, M. A. (1960). "The effect of bilateral ablation of the amygdala on endocrine function in the rat," Endocrinology, 66:565-574.

Yates, F. E., Russell, S M., and Maran, J W. (1971). "Brain-adrenohypophysial communication in mammals," Ann. Rev. Physiol., 33:393-444.

Yeh, S. D. J., Solomon, J. D., and Chow, B. F. (1959). "Influence of vitamin B_6 on tissue serotonin levels in rats," Federat. Proc., 18:357.

Yoshio, H. (1964). "Inhibitory effect of preoptic stimulation on adrenal 17-hydroxycorticosteroid secretion rate in the cat," Nature, 201:1334-1335.

Yudaev, N. A., and Afinogenova, S. A. (1962). "The state of the pituitary-adrenocortical system after administration of cortisone and ACTH under chronic experimental conditions," Probl. Éndrokrinol. i Gormonoter., 8(2):12-20.

Yudaev, N. A., and Pankov, Yu. A. (1958). "A modification of the method of Silber and Porter for estimation of 17-hydroxycorticosteroid in the peripheral blood plasma," Probl. Éndokrinol. i Gormonoter., 4(2):35-42.

Yuwiler, A., Geller, E., Schapiro, Sh., and Slater, G. (1965). "Adrenocortical and enzymic effects of imipramine and chlorpromazine," Biochem. Pharmacol., 14:621-623.

Zaitsev, V. F. (1959). "Changes in the blood white cell count and formula after a single injection of ACTH," Sov. Meditsina, (11):35-42.

Zarrow, M. X., Denison, M. E., Rosenberg, B., Mann, D. E., and Neher, G. M. (1952). "Effect of insulin and epinephrine on the eosinophil and blood glucose levels in sheep; lack of diurnal rhythm," Amer. J. Physiol., 171:636-640.

Zeimal, E. V., Mikhel'son, M. Ya., and Rybolovlev, R. S. (1957). "Connection between the chemical structure and pharmacological action of some cholinolytic, cholinomimetic, and anticholinesterase drugs," in: The Physiological Role of Acetylcholine and the Search for New Drugs [in Russian], Leningrad, pp. 424-441.

Zeimal, E. V., and Rybolovlev, R. S. (1957). "The central and peripheral action of some cholinomimetics containing tertiary nitrogen and their iodomethylates," in: The Physiological Role of Acetylcholine and the Search for New Drugs [in Russian], Leningrad, pp. 338-354.

Zeller, E. A., and Barsky, J. (1952a). "*In vivo* inhibition of liver and brain monoamine oxidase by 1-isonicotinyl-2-isopropylhydrazine," Proc. Soc. Exptl. Biol. (New York), 81:459-461.

Zeller, E. A., Barsky, J., Berman, E.R., and Fouts, J.R. (1952b). "Action of isonicotinic acid hydrazide and related compounds on enzymes of brain and other tissues," J. Lab. Clin. Med., 40:965-966.

Zeller, E. A., Barsky, J., Berman, E. R., and Fouts, J. R. (1952c). "Action of isonicotonic acid hydrazide and related compounds on enzymes involved in the autonomic nervous system," J. Pharmacol. (Baltimore), 106:427-428.

Zeller, E. A., Barksy, J., and Berman, E. R. (1955). "Amine oxidases, XV: inhibition of monoamine oxidase by 1-isonicotinyl 1-2-isopropylhydrazine," J. Biol. Chem., 214:267-274.

Zeman, W., and Innes, J.R. M. (1963). Craigie's Neuroanatomy of the Rat, Academic Press, New York and London.

Zoryan, V. G. (1965). "The effect of some monoamine oxidase inhibitors on the pituitary-adrenocortical system," Farmakol. i Toksikol., 28(4):409-412.

GPSR Compliance

The European Union's (EU) General Product Safety Regulation (GPSR) is a set of rules that requires consumer products to be safe and our obligations to ensure this.

If you have any concerns about our products, you can contact us on ProductSafety@springernature.com

In case Publisher is established outside the EU, the EU authorized representative is:

Springer Nature Customer Service Center GmbH
Europaplatz 3
69115 Heidelberg, Germany

Batch number: 10620163

Printed by Printforce, the Netherlands